Electrostatic Lens Systems

Advisory Panel for Adam Hilger Books on Atomic and Molecular Physics

ELECTROSTATIC LENS SYSTEMS

Professor D W O Heddle

Royal Holloway and Bedford New College
University of London

Adam Hilger
Bristol, Philadelphia and New York

© IOP Publishing Ltd 1991

British Library Cataloguing in Publication Data

Heddle, Prof D. W. O.
 Electrostatic lens systems. - (Adam Hilger series
 on atomic and molecular physics)
 I. Title II. Series
 535

 ISBN 0-7503-0119-8

Library of Congress Cataloging-in-Publication Data

Heddle, D. W. O. (Douglas W. O.)
 Electrostatic lens systems / D.W.O. Heddle.
  p. cm.
 Includes bibliographical references.
 ISBN 0-7503-0119-8
 1. Electrostatic lenses. 2. Electrostatic lenses—Design and
construction—Data processing. I. Title.
QC544.E5H43 1991
5371.2—dc20   91-11678
                                                        CIP

Published under the Adam Hilger imprint by IOP Publishing Ltd
Techno House, Redcliffe Way, Bristol BS1 6NX, England
335 East 45th Street, New York, NY 10017-3483, USA
US Editorial Office: 1411 Walnut Street, Suite 200, Philadelphia, PA 19102

Printed in Great Britain by J W Arrowsmith Ltd, Bristol

Contents

Introduction

Interest in electrostatic lenses for the control of ion and (especially) electron beams is of long standing and has grown considerably in the past few decades. In addition, recent innovations in the production of low energy positrons have opened a whole new field of research for which electrostatic lenses are required. This book makes no attempt to be a comprehensive text on electron and ion optics, but addresses the need for data on simple lenses consisting of two or three apertures or cylinders, together with some more complex cylinder lenses which have interesting properties. There are a number of quite detailed and extensive texts, of which two recent books, by Hawkes and Kasper [1] and by Szilagyi [2], contain modern treatments of the subject and are well worth consulting. The book by Harting and Read [3] which has objectives similar to the present volume has, unfortunately, been out of print for some years.

The book itself forms an *introductory* text on electrostatic optics and is accompanied by a disc containing a program for the IBM personal computer. This program, which is described in detail in Chapter 6 and in the Appendix, will calculate the focal and aberration properties of a range of lenses and will also allow the study of the imaging behaviour between conjugate planes. Ray paths are traced in the paraxial approximation. A graphical display (Hercules, CGA, EGA or VGA) is required to run the program, and a maths coprocessor is very desirable. Only non-relativistic energies are considered and the electrode geometries are all very simple and easily manufactured.

The text bears some resemblance to courses I have given to first year postgraduate students over a number of years. The **LENSYS** program, on the other hand, is the result of a pressing need for lens data in a readily usable form for research purposes, and includes a number of lenses which have been found to be particularly useful for the control of electron and positron beams.

It is a pleasure to acknowledge the contribution made by two former research students, Dr Roger Cook and Dr Tony Renau, to the development of the Bessel function expansion method and its application to reveal hitherto unsuspected interrelationships between the aberration coefficients. I am particularly indebted to my colleague, Dr Susan Kay, who has worked with the **LENSYS** program from its earliest manifestation as a number of separate routines. Her suggestions of material to include or omit, and her assistance in developing a reasonably 'user-friendly' interface have been invaluable.

Chapter 1

The Optics of Simple Lenses

1.1 Analogies between particle and photon optics

The action of lenses and mirrors in controlling beams of light are matters of everyday experience and it may be helpful to look at some systems which have similar behaviour in photon and particle optics. The simplest, shown in figure 1.1, is the plane boundary separating two regions which differ in some property. In the case of photon optics the important property is the refractive index.

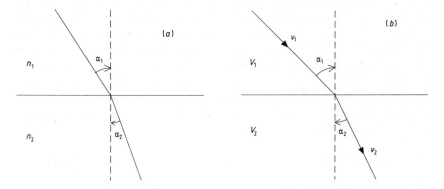

Figure 1.1 (a)Refraction of light at a plane boundary between two media having refractive indices n_1 and n_2; and (b) deviation of a beam of charged particles at a plane boundary seperating regions having potentials V_1 and V_2.

The path of a ray of light incident non-normally onto the boundary is changed on crossing the boundary, the directions in the two regions being related by Snell's law. The analogue in particle optics is a boundary separating two regions at different electrostatic potentials. Here we consider a particle having a velocity v_1 in the first region and v_2 in the second. These velocities are related to the potentials, V_1 and V_2, in the two regions by

$$\tfrac{1}{2}mv_1^2 + eV_1 = \tfrac{1}{2}mv_2^2 + eV_2 = 0 \qquad (1.1)$$

where e is the particle charge. This expression, which is fundamental to the simple analysis of electrostatic lenses, defines the zero of potential as that for which the particle is at rest. The potential changes abruptly at the boundary, but there is no change parallel to the boundary and so no force parallel to the boundary acts on the particle. The component of momentum parallel to the boundary is therefore unchanged and we can write

$$mv_1 \sin \alpha_1 = mv_2 \sin \alpha_2 \qquad (1.2)$$

where α_1 and α_2 are the angles between the normal to the boundary and the path of the particle in the two regions. From equations (1.1) and (1.2) we have $\sin \alpha_1 / \sin \alpha_2 = (V_2/V_1)^{1/2}$ which is exactly Snell's law with $V^{1/2}$ playing the role of the refractive index.

That example was rather artificial, because abrupt changes of potential do not occur in free space, but can only be approximated by the use of closely spaced grids. Our second example is more realistic and demonstrates the focusing behaviour of a lens consequent on *curved* boundaries. Figure 1.2(a) depicts two regions, having refractive indices n_1 and n_2, separated by a spherical boundary of radius R. Rays of light incident parallel to the axis, and at small distances h_2 and h_1 from the right and left, are refracted at the boundary in such a way that they cross the axis at points labelled F_1 and F_2, respectively. These are the first and second focal points of this simple lens and the distances measured from the boundary (and here we assume that $h \ll R$ so that we can define the position of the boundary by the pole, O, where the boundary cuts the axis) are called the first and second focal lengths and denoted by f_1 and f_2, respectively. Remembering that h/R is very small such that the sines and tangents of the angles can be approximated by the angles themselves, we can then write for the upper part of figure 1.2(a)

$$n_1\alpha_1 = n_2\alpha_2 \qquad\qquad \alpha_1 = h_1/R \qquad\qquad \alpha_1 - \alpha_2 = h_1/f_2$$

and for the lower part

$$n_1\alpha_4 = n_2\alpha_3 \qquad\qquad \alpha_3 = h_2/R \qquad\qquad \alpha_3 - \alpha_4 = h_2/f_1.$$

Eliminating the angles we find

$$\frac{1}{f_1} = \frac{(n_1 - n_2)}{n_1}\frac{1}{R} \qquad\qquad\qquad \frac{1}{f_2} = \frac{(n_2 - n_1)}{n_2}\frac{1}{R}.$$

Taking the ratio of these two expressions we see that the ratio of the two focal lengths is the negative of the ratio of the refractive indices. The sign is a consequence of defining the focal lengths as distances measured *from* the boundary and taking the pole to be the origin of cartesian coordinates.

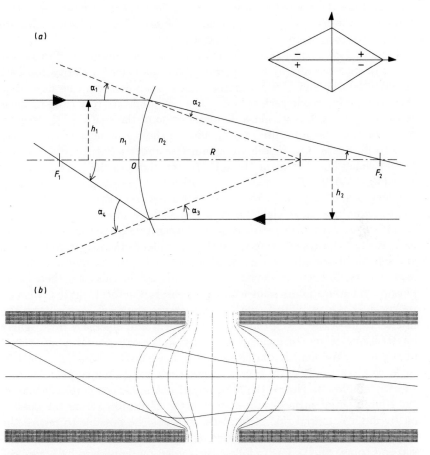

Figure 1.2 (*a*) Refraction of light at a spherical boundary between two media having refractive indices n_1 and n_2. The inset illustrates the sign convention. The angles α_1 and α_2 are positive as are the angles between the rays and the axis at the focal points. Since α_3 and α_4 are negative, the angle at F_1 is equal to $\alpha_3 - \alpha_4$. (*b*) Paths of charged particles in a two-cylinder lens. The paths become asymptotic to straight lines only away from the central region of the lens. The values of the equipotential surfaces, shown in section, are discussed in the text.

The sign convention for angles is that angles measured from the axis to the ray are positive if the rotation is clockwise. These conventions lead to angles having the signs of their tangents, with the two distances involved measured *from* the right angle. One consequence of this choice is that the sign of the angle a ray makes with the axis is opposite to that of dr/dz for that ray. For angles defined with respect to some other direction, such as a radius (as in figure 1.2(*a*)) or a second ray (as in figure 1.3), clockwise rotation from the ray to the other direction defines the positive sense. In a number of the figures we have indicated the direction in which angles are measured according to these conventions.

Figure 1.2(b) illustrates an electrostatic analogue consisting of two coaxial cylinders separated by half their common diameter for clarity and held at potentials V_1 and V_2. No mathematical analysis should be needed to show that, deep inside each cylinder, the potentials are very close to those of the electrodes, but within a diameter or so of the centre of the system the potential changes quite rapidly and we show a number of equipotentials to illustrate this. These are symmetric about a flat, central, equipotential which has a value of $(V_2 + V_1)/2$ and those to the left have values $V_1 + (V_2 - V_1)/2^n$ where $n = 2 \ldots 5$. Notice that the paths of the particles are curved and only away from the central region of the lens do they become asymptotic to straight lines. We shall have more to say about the curvature later, but for the moment we concentrate on the asymptotes. As in figure 1.2(a), rays entering the system parallel to the axis from the right and left are deviated towards the axis, crossing it at the first and second focal points, but as the deviation is not abrupt in this case there is no *physical* reference point from which to measure the focal lengths. Instead we consider the asymptotic paths and measure the focal lengths from the intersections of the asymptotes to the incident and emergent rays. These intersections define, for incident rays close to the axis, the *principal planes* of the lens. We shall see later that, for non-paraxial rays, the principal surfaces are actually curved; nonetheless the term 'principal plane' is universally employed. Notice that the actual paths are not perpendicular to the equipotential surfaces and approach the asymptotes from the outside on the high potential side of the lens. It would be possible to regard the equipotentials as boundaries separating regions having some mean potential and to use Snell's law to follow the refraction, boundary by boundary, through the lens, but this would be rather tedious and not very accurate. There are much better methods.

1.2   The cardinal points of a lens

The contrast between the ray paths in the photon and particle lenses of the previous section was very marked. In the case of a photon lens of many elements, such as is common in even quite simple cameras, the ray paths, while still made up of straight line segments, may experience many changes of direction within the lens and resemble somewhat the ray paths in the particle lens. The details of the paths are of little importance as long as the asymptotic paths are well defined and the appropriate planes of intersection known, because the imaging properties of both sorts of lens can be completely described (in the paraxial approximation, at least) by the positions of the focal points and the principal points, that is, the intersections of the principal planes with the axis. These are four of the six *cardinal points* of the lens and their use is illustrated in figure 1.3. The lens is specified in this figure by these four points, and their separations and the distances from an arbitrary reference plane, R, are indicated. If there is a plane of (mechanical) symmetry in the lens, this plane is commonly used as reference and the distances from this plane to the focal points are called the 'mid-focal distances' and are usually denoted by F_1 and F_2. The ambiguity of the same symbols being used for these distances and for the focal points themselves is not usually a problem.

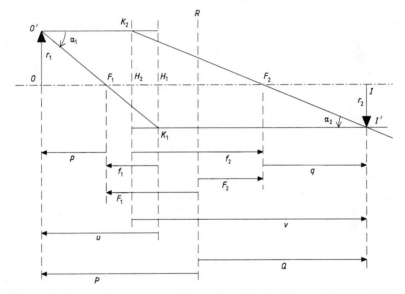

Figure 1.3   This diagram shows the focal and principal points of a lens and illustrates a construction, using asymptotes to the principal rays, for finding the position of the image of a given object. The angles α_1 and α_2 are measured from the upper ray to the lower and have opposite signs.

Chapter 5 will contain a discussion of the aberrations of particle lenses, but until then we shall concentrate on the rather simpler situation of Gaussian imaging appropriate to paraxial rays, and in the next section shall develop some important interrelations. We have restricted the field of this book to lenses of axial symmetry, and we shall consider only rays which lie in planes containing the optic axis, that is, the meridional rays.

1.2.1   The construction of an image

We consider an object, OO' of height r_1 at a distance p from the first focal point, u from the first principal point or P from the reference plane, and construct the image II' by drawing two *principal rays* from O'. The first is drawn parallel to the axis to represent the initial asymptote and, from its intersection with the second principal plane at K_2, a line representing the asymptote to the emergent ray is drawn to pass through the second focal point, F_2. The second is drawn through the first focal point to meet the first principal plane at K_1 from where it continues parallel to the axis. The intersection of these two rays defines the image point, I'. The image has a height r_2 and is a distance q from the second focal point, v from the second principal point or Q from the reference plane. With our cartesian sign convention p, P, f_1, F_1, u and r_2 are all negative. The other distances are positive. The angles α_1 and α_2 have opposite signs because the rotations from one ray to the other are in different senses: we have chosen to mark α_1 as the positive angle.

In order to find relationships among the various distances, we consider a number of similar triangles. From triangles OF_1O' and $H_1F_1K_1$ we see that $r_1/p = -r_2/f_1$ and, from triangles IF_2I' and $H_2F_2K_2$, $r_2/q = -r_1/f_2$. Triangles $H_1F_1K_1$ and $H_2F_2K_2$ are not similar and allow us to represent the angles as $\alpha_1 = r_2/f_1$ and $\alpha_2 = -(-r_1/f_2)$. The (transverse) magnification of the lens, M, is given by the ratio r_2/r_1 and the angular magnification, M_α, by α_2/α_1 so we can write

$$M = -f_1/p = -q/f_2 \qquad \text{and} \qquad M_\alpha = -(r_1/r_2)(f_1/f_2)$$

so

$$M M_\alpha = -f_1/f_2 = n_1/n_2 = (V_1/V_2)^{1/2}. \tag{1.3}$$

This last relationship, between the magnification, angular magnification and the ratio of the potentials in the object and image regions, is known as the law of Helmholtz and Lagrange and is essentially a statement that the brightness of an image cannot exceed that of the object.

Using the two expressions for the magnification we obtain $pq = f_1f_2$, which is Newton's relation, and if we substitute for u $(= p + f_1)$ and v $(= q+f_2)$ we obtain $f_1/u + f_2/v = 1$. This is not often used (except in an approximate form valid for thin lenses), but substitution for P $(= p + F_1)$ and Q $(= q + F_2)$ shows that a graph of Q against P is a rectangular hyperbola with its centre at F_1, F_2. This is frequently used to present data for lenses of two elements with the voltage ratio as a parameter. Lines of constant magnification can be shown on the same axes, leading to a very simple procedure for the choice of a (two-element) lens to meet a specification in terms of a voltage ratio and magnification. In the same way that it is only the *ratio* of the potentials which govern the behaviour of a lens, so the actual size of the lens does not matter and it is only the ratios P/D, F_1/D, etc, where D is some characteristic dimension of the lens, usually the diameter, which are important. It is very common indeed to find focal lengths and other distances represented with D as the unit of length. In this book we shall normally write 'F_1', for example, instead of 'F_1/D' unless a specific point is to be made. Figure 1.4 illustrates the general form of a family of P–Q curves showing both branches of the hyperbolae and also the real object–image region for a lens of two closely spaced coaxial cylinders.

With our sign convention, the transverse and angular magnifications of a real image are negative. It is, however, common usage to speak of magnifications as though they were positive and to ignore the sign when using terms such as 'larger', 'smaller', 'increase', etc. Except in critical situations we shall cite magnifications in this fashion.

It is easy to see that an object in the first principal plane will be imaged in the second principal plane unchanged in size and the principal planes are sometimes referred to as 'planes of unit *positive* magnification'. Note that either the object or the image must be virtual in this case. We shall see later that this property of the principal planes has been exploited in particle optics.

1.2.2 The general ray

While the two principal rays shown in figure 1.3 offer a simple construction of the position and size of an image, a ray from an axial object point to the corresponding image point is often of greater interest. Provided that we restrict ourselves to paraxial, Gaussian imaging, any linear combination of two rays which themselves satisfy the focusing condition will also satisfy that condition. Figure 1.5 illustrates a construction to demonstrate this property using the asymptotes to the various rays. Any independent pair of rays may be used as the basis, but it is convenient to choose rays which are parallel to the axis in one region. We define the first ray, $r_1(z)$, as that passing through the first focal point of the lens at an angle of $+45°$. This ray is therefore parallel to the axis in the image space and a distance f_1 from the axis. The second ray, $r_2(z)$, is parallel to the axis in the object space and a distance, $-f_1$, from the axis. It crosses the axis at the second focal point, with a slope f_1/f_2.

The general ray, $r(z)$, is then represented by $r(z) = \zeta r_1(z) + \xi r_2(z)$. The object is a distance p from the first focal point, so at the object we have $r(z) = \zeta(-p) + \xi(-f_1) = 0$, and we can therefore write $\xi = \zeta/M$ where M is the lens magnification between these conjugates. At the image we write $r(z) = \zeta[f_1 + q(f_1/f_2)M^{-1}] = 0$ for all values of ζ because the term in the square brackets is zero. The single parameter, ζ, is sufficient to define a particular ray and we write

$$r(z) = \zeta \left(r_1(z) + \frac{1}{M} r_2(z) \right).$$

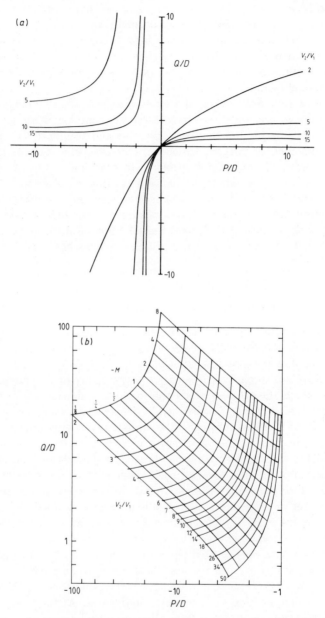

Figure 1.4 (a) *P–Q* diagram for a lens. The upper left region ($P < 0$, $Q > 0$) applies to the most common case of *real* objects and images. In the other regions one or both are virtual. Notice that, close to the origin, the *P–Q* relationship is virtually independent of the voltage ratio. (b) *P–Q* diagram for a two-cylinder lens, with logarithmic scales to simplify the presentation of data over a wide range of distances and voltage ratios. Lines of constant magnification and of constant voltage ratio are shown.

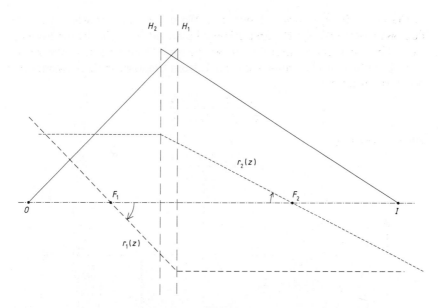

Figure 1.5   Asymptotes to two principal rays, $r_1(z)$ and $r_2(z)$, and to a general ray $r(z) = -[r_1(z) + (1/M)r_2(z)]$.

Figure 1.5 illustrates the construction of the asymptotes to one such general ray. We take $\zeta = -1$, and draw the asymptote from the object point, O, until it reaches the first principal plane, H_1. Then, from the second principal plane, H_2, at the same radial distance, draw the emergent asymptote to the image point. The slope of this asymptote is $(-1/M)(f_1/f_2)$ and, as M is known from the ratio of the first focal length and the object distance, the position of the image point does not have to be known beforehand. The values used for illustration in figure 1.5 are

$$f_1 = -4 \qquad f_2 = 8 \qquad p = -5 \qquad q = 6.4$$

giving

$$M = -0.8$$

1.2.3   The nodal points

The two remaining cardinal points are the *nodal points*, N_1 and N_2. These are points of unit positive *angular* magnification in the sense that a ray directed towards the first emerges as though from the second with the same slope. All rays from a point on the object must pass through the corresponding point on the image and we show a particular linear combination of the two principal rays in figure 1.6. We draw a ray for which the incident asymptote crosses the first principal plane at some distance h from the axis. The emergent asymptote appears to come from a point in the second principal plane at this same distance, h, from the axis. If these asymptotes are parallel we have

$$\frac{r_1 - h}{p + f_1} = \frac{r_2 - h}{q + f_2}$$

which leads to

$$h = r_1 \left(\frac{f_1 + f_2}{f_2 - p} \right).$$

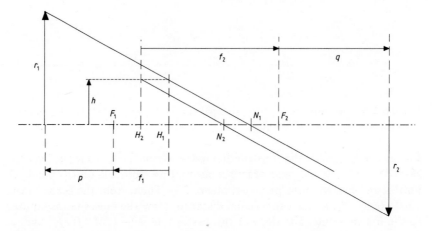

Figure 1.6   Diagram to illustrate the action at the nodal points of a lens.

These asymptotes cross the axis at the nodal points which are clearly separated by the same distance as the principal points. We leave it as an exercise for the reader to show that these points are at distances f_2 and f_1 from the focal points F_1 and F_2, respectively. Notice that if the refractive indices (potentials) on the two sides of the lens are the same, i.e. $f_1 = -f_2$, then $h = 0$ and the asymptotes cross the axis at the principal points which, for this case, coincide with the nodal points.

1.3   Matrix representation of lens parameters

While the focal lengths and distances are the parameters most often presented, it is possible to express the action of a lens by a 2 by 2 matrix which acts on some initial ray asymptote for which the ray position, r, and slope, r', at some plane are expressed as a column vector, to produce a modified asymptote at a second plane.

$$\begin{pmatrix} r_2 \\ r'_2 \end{pmatrix} = \begin{pmatrix} a_{11} & a_{12} \\ a_{21} & a_{22} \end{pmatrix} \begin{pmatrix} r_1 \\ r'_1 \end{pmatrix}. \tag{1.4}$$

The planes between which the matrix acts determine the values of the matrix elements and some compromise is needed between simple expressions for the elements and a convenient choice of planes. The significance of the individual elements is clear from equation (1.4): a_{11} is the linear magnification between the planes and a_{22} the angular magnification. The other elements relate the radius at one plane and the slope at the other. The simplest matrix is one which transfers the ray from the first principal plane to the second. In this case the element $a_{11} = 1$ since the principal planes are planes of unit positive magnification, and $a_{22} = -f_1/f_2$ from equation (1.3). r_2 does not depend on r'_1 so $a_{12} = 0$ and it is easy to show that $a_{21} = -1/f_2$ by considering an incident asymptote parallel to the axis.

The positions of the principal planes of an electrostatic lens vary with the potentials and it would be more convenient to work with a matrix operating between planes which did not move in this way. We now develop the matrix which transfers an incident asymptote at the reference plane to an emergent asymptote *at the same plane.* Despite the change in radial position which may occur, this is usually referred to as a *bending matrix.* We do this by operating on the incident column vector with three matrices. The first transfers the ray from the reference plane to the first principal plane, the second is the matrix acting between the principal planes and the third transfers the ray from the second principal plane to the reference plane.

$$\begin{pmatrix} r_2 \\ r'_2 \end{pmatrix} = \begin{pmatrix} 1 & f_2 - F_2 \\ 0 & 1 \end{pmatrix} \begin{pmatrix} 1 & 0 \\ -\frac{1}{f_2} & -\frac{f_1}{f_2} \end{pmatrix} \begin{pmatrix} 1 & F_1 - f_1 \\ 0 & 1 \end{pmatrix} \begin{pmatrix} r_1 \\ r'_1 \end{pmatrix}$$

$$= \begin{pmatrix} \frac{F_2}{f_2} & \frac{F_1 F_2 - f_1 f_2}{f_2} \\ -\frac{1}{f_2} & -\frac{F_1}{f_2} \end{pmatrix} \begin{pmatrix} r_1 \\ r'_1 \end{pmatrix}.$$

To use this matrix to study the formation of an image we first translate the ray from the object to the reference plane, a distance $-P$, then use the lens matrix to produce the emergent ray and finally transfer this ray a distance L. The product of these three matrices gives the *transfer matrix*

$$\begin{pmatrix} 1 & L \\ 0 & 1 \end{pmatrix} \begin{pmatrix} \frac{F_2}{f_2} & \frac{F_1 F_2 - f_1 f_2}{f_2} \\ -\frac{1}{f_2} & -\frac{F_1}{f_2} \end{pmatrix} \begin{pmatrix} 1 & -P \\ 0 & 1 \end{pmatrix} = \begin{pmatrix} \frac{F_2 - L}{f_2} & \frac{(F_2 - L)(F_1 - P) - f_1 f_2}{f_2} \\ -\frac{1}{f_2} & \frac{P - F_1}{f_2} \end{pmatrix}$$

The condition for imaging is that the *position* of the ray in the exit plane does not depend on the *slope* of the incident ray. This is equivalent to the element, a_{12}, of the transfer matrix being zero, which will be the case if $L = Q$. The element a_{11} is equal to the magnification of the lens between the conjugate planes, and a_{22} is the angular magnification.

The analysis of combinations of lenses is frequently simpler if the matrix formulation is used, because the only calculation required is a sequence of matrix multiplications. A further situation in which the matrix formulation has advantages is the analysis of very weak two-element lenses. This is because the matrix elements vary smoothly as the potential ratio passes through one, in contrast to the focal lengths and distances which become infinite for this value of the ratio. Figure 1.7 shows the matrix elements for a lens of two coaxial cylinders separated by a gap of 0.1 diameters.

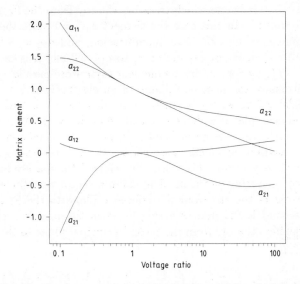

Figure 1.7 Matrix elements for a lens of two coaxial cylinders separated by 0.1 diameters. Note that all four matrix elements are continuous as the potential ratio passes through a value of one.

Chapter 2

The Motion of Charged Particles in an Electrostatic Field

2.1 The equation of paraxial motion

Our concern is with axisymmetric lenses and so we use cylindrical polar coordinates with the z axis as the axis of symmetry of the lens system. Laplace's equation is

$$\frac{\partial^2 V}{\partial z^2} + \frac{1}{r}\frac{\partial}{\partial r}\left(r\frac{\partial V}{\partial r}\right) = 0 \tag{2.1}$$

where $V(r, z)$ is the potential. Because of the rotational symmetry an expansion of the potential in *even* powers of r can be made

$$V(r, z) = \sum_{n=0}^{\infty} A_n(z) r^{2n}.$$

The two terms of Laplace's equation are

$$\frac{1}{r}\frac{\partial}{\partial r}\left(r\frac{\partial V}{\partial r}\right) = \sum_{n=0}^{\infty} 4n^2 A_n(z) r^{2n-2}$$

and

$$\frac{\partial^2 V}{\partial z^2} = \sum_{n=0}^{\infty} A_n''(z) r^{2n}$$

where primes (′) are used to denote differentiation with respect to z. The sum of the coefficients of each power of r must be zero so we have a recurrence relation

$$A_{n+1}(z) = -\frac{A_n''(z)}{4(n+1)^2}$$

giving the series expansion as

$$V(r,z) = A_0(z) - \frac{A_0''(z)r^2}{2^2} + \frac{A_0^{(4)}(z)r^4}{2^2 \cdot 4^2} + \cdots. \tag{2.2}$$

The axial potential, $V(z)$, is just $A_0(z)$ and we can write the axial and radial components of the electric field as

$$E_z = -\frac{\partial V}{\partial z} = -V' \qquad\qquad E_r = -\frac{\partial V}{\partial r} = \frac{r}{2}V''$$

where we omit the explicit reference to z and retain terms only to second order since we are considering only paraxial rays. When we come to consider the aberrations of particle lenses we shall have to make allowance for the higher-order terms.

The axial velocity is so much greater than the radial in these circumstances that we can write $\frac{1}{2}m(dz/dt)^2 + eV = 0$ for the total energy of the particle. The equation of radial motion is

$$m\frac{d^2r}{dt^2} = eE_r = \frac{er}{2}V''$$

writing e for the charge on the particle. The sign of the charge matters only insofar as the potential must always have the opposite sign and we shall soon see that the magnitude of the charge does not matter. To eliminate t from these expressions we write

$$\frac{d^2r}{dt^2} = \frac{dz}{dt}\frac{d}{dz}\left(\frac{dz}{dt}\frac{dr}{dz}\right)$$

giving

$$\left(\frac{-2eV}{m}\right)^{1/2}\frac{d}{dz}\left[\left(\frac{-2eV}{m}\right)^{1/2}\frac{dr}{dz}\right] = \frac{er}{2m}V''$$

which reduces to

$$\frac{d^2r}{dz^2} + \frac{1}{2}\frac{V'}{V}\frac{dr}{dz} = -\frac{r}{4}\frac{V''}{V}. \tag{2.3}$$

This equation contains neither the charge of the particle nor its mass and it is therefore valid for electrons, positrons and ions of either sign. The only constraint is that the potential should have a sign opposite from that of the particle to ensure a positive total energy.

In order to integrate this equation of motion we require very precise information about the axial potential, because we need to determine the second derivative. An alternative approach is to change the independent variable in such a way as to remove this term. We introduce a *reduced radius*, R, defined by $R = rV^{1/4}$ where V must be read as $|V|$ for positive particles. Successive differentiation of this expression yields

$$\frac{\mathrm{d}R}{\mathrm{d}z} = V^{1/4}\frac{\mathrm{d}r}{\mathrm{d}z} + \frac{1}{4}V^{-3/4}V'r$$

$$\frac{\mathrm{d}^2R}{\mathrm{d}z^2} = V^{1/4}\left[\underbrace{\frac{\mathrm{d}^2r}{\mathrm{d}z^2} + \frac{1}{2}\frac{V'}{V}\frac{\mathrm{d}r}{\mathrm{d}z} + \frac{r}{4}\frac{V''}{V}}_{=\,0} - \frac{3r}{16}\left(\frac{V'}{V}\right)^2\right]$$

where equation (2.3) shows that the grouped terms sum to zero, and we are left with the very simple equation of motion

$$\frac{\mathrm{d}^2R}{\mathrm{d}z^2} = -\frac{3}{16}R\left(\frac{V'}{V}\right)^2 \tag{2.4}$$

which is known as the Picht equation [4].

2.2   Some general results

The independent variable in the Picht equation is the ratio of the axial potential gradient to the potential itself and it will be convenient to define a new variable, $T(z) = V'(z)/V(z)$, and to note that, because this appears *squared* the sign does not matter. The general form of $T(z)$ is illustrated for a lens with two cylindrical electrodes in figure 2.1(a). The maximum value of $|T|$, and therefore the main focusing action of the lens, occurs on the low potential side, showing that the principal planes will be found on that side. Figure 2.1(b) shows rays traced in such a lens by the **LENSYS** program.

It is not difficult to show that all electrostatic lenses having uniform potential regions to each side are converging and to obtain an approximate expression for their focal lengths. We consider a ray incident parallel to the axis at a reduced radius R_1 and make the assumption that this reduced radius does not change in passing through the lens. Naturally, r, the true radius, will change or there would be no lens action, but the change in V will act in the opposite sense. A formal integration of the Picht equation then gives

$$\int_{-\infty}^{\infty} R''\mathrm{d}z = R_2' - R_1' = -\frac{3}{16}R_1\int_{-\infty}^{\infty} T^2\mathrm{d}z$$

though in practice the limits of integration can be very much narrower as a consequence of the sharply peaked nature of T^2. $R_1' = 0$ since the incident ray is parallel to the axis and so R_2' has the opposite sign to R_1. Writing

$$R_2' = r_2'V_2^{1/4} + \frac{r_2}{4}V_2^{-3/4}V_2'$$

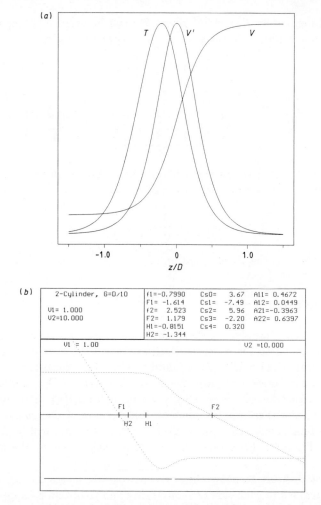

Figure 2.1 (*a*) The variation of $V(z)$, $V'(z)$ and their ratio, $T(z)$ with axial position z/D in a two-cylinder lens. Note that the maximum value of T occurs on the low potential side of the lens. The ordinate scale is arbitrary. (*b*) Ray paths in this lens showing that most of the deviation of the particle paths occurs in the region where $|T|$ is large. The focal and principal points of the lens are also shown.

and, noting that V_2' will be zero away from the lens proper and that $V_2^{1/4}$ is intrinsically positive, we see that r_2' has the same sign as R_2' and so for rays incident above the axis the emergent ray moves towards the axis giving a *convergent* lens action. Lenses for which the object and image positions lie in regions of uniform potential are known as *immersion lenses*.

We saw in section 1.2.1 that the second focal length of a lens can be written as $-r_1/\alpha_2$. We identify α_2 with r_2' and write

$$\frac{1}{f_2} = -\frac{r_2'}{r_1} = -\frac{R_2'}{R_1}\left(\frac{V_1}{V_2}\right)^{1/4} = \frac{3}{16}\left(\frac{V_1}{V_2}\right)^{1/4}\int_{-\infty}^{\infty}\left(\frac{V'}{V}\right)^2 dz. \quad (2.5)$$

If we were to trace a ray incident parallel to the axis, but from the other side, we would obtain

$$\frac{1}{f_1} = -\frac{r_1'}{r_2} = -\frac{R_1'}{R_2}\left(\frac{V_2}{V_1}\right)^{1/4} = \frac{3}{16}\left(\frac{V_2}{V_1}\right)^{1/4}\int_{\infty}^{-\infty}\left(\frac{V'}{V}\right)^2 dz \quad (2.6)$$

and would find, for the ratio of the focal lengths,

$$\left(\frac{f_1}{f_2}\right) = -\left(\frac{V_1}{V_2}\right)^{1/2} = -\frac{n_1}{n_2}.$$

The ratio is negative because the integrals in equations (2.5) and (2.6) have opposite signs and the result is what we expect from our previous general discussion.

An alternative expression is sometimes quoted

$$\frac{1}{f_2} = \frac{1}{8V_2^{1/2}}\int_{-\infty}^{\infty}\frac{V'^2}{V^{3/2}}dz$$

but the derivation of this expression is valid only if V'' does not change sign and this is never true unless the potential is constrained by grids.

2.2.1 The nodal and principal points

We will now consider an incident ray having its asymptote directed towards the first nodal point as illustrated in figure 2.2. By a suitable choice of vertical scale the same line can be used to represent the asymptotes of both the ray itself and the reduced ray. The Picht equation shows that $R'' \propto -R$ and so the path of the reduced ray is always curved *towards* the axis. The reduced ray therefore crosses the axis to the left of the first nodal point; the curvature then reverses and the ray approaches its emergent asymptote from below. This asymptote passes through the second nodal point which must lie to the left of the point at which the ray crossed the axis. Note that both the 'real' ray (shown by the dotted line) and the reduced ray cross the axis at the same point. While the slopes of the real asymptotes are equal, those of the reduced asymptotes are in the ratio of the one fourth power of the potential ratio. As figure 2.2 shows, the actual ray path is not bounded by the asymptotes. In Chapter 1 we saw that the separation

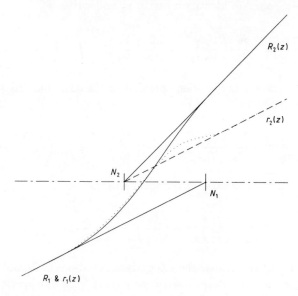

Figure 2.2 Paths of the ray $r(z)$ ($\cdots$) and the reduced ray $R(z)$ (——) in the vicinity of the nodal points. The reduced ray is bounded by the asymptotes and the curvature of its path is always towards the axis. The slopes of the asymptotes to the ray, $r(z)$, are equal, while those of the reduced ray depend on the potential ratio. The radial scales have been chosen to make the incident asymptotes coincide.

of the nodal points is equal to that of the principal points and our result, above, shows that the principal points of an electrostatic immersion lens are crossed, in the sense that H_2 is further to the low potential side of the lens than H_1. This is just the way in which the principal planes were represented in figure 1.3.

2.2.2 Relativistic effects

The independence of the paths of charged particles in electrostatic fields from the charge to mass ratio is only true for velocities sufficiently small that the mass of the particle remains essentially equal to the rest mass. Certainly the lenses discussed in Chapter 6 and illustrated in the program would not be suitable for use with high potentials, but it is perhaps worth noting here the ways in which relativistic effects can be taken into account. Writing the momentum of a particle as

$$p = m_0 v (1 - \beta^2)^{1/2} \tag{2.7}$$

and the kinetic energy as

$$E = m_0 c^2 ((1 - \beta^2)^{-1/2} - 1) \qquad (2.8)$$

the velocity can be expressed in terms of the momentum (using equation (2.7)) and substituted into equation (2.8) to give

$$m_0 c^2 + E = c\sqrt{m_0^2 c^2 + p^2}$$

which can be rearranged to give an explicit expression for the momentum in terms of the kinetic energy. Recalling that this is just equal to the negative of the charge multiplied by the potential (the zero being taken as usual for the particle to be at rest), we finally express the momentum in terms of the accelerating potential, V.

$$p^2 = 2m_0 E \left(1 + \frac{E}{2m_0 c^2}\right)$$

$$= -2m_0 eV \left(1 - \frac{eV}{2m_0 c^2}\right)$$

$$= -2m_0 eV_{rel}$$

where

$$V_{rel} = V \left(1 - \frac{eV}{2m_0 c^2}\right) = V(1 + \epsilon)$$

$$= V(1 + 0.987 \times 10^{-6} V) \qquad \text{for electrons}$$

is known as the relativistically corrected acceleration potential. Note that the product eV is always negative, and therefore ϵ is a positive quantity.

The equations of motion can be analysed in terms of this corrected potential and the Picht equation retains much the same form. The reduced radius, R, is now defined by $R = rV_{rel}^{1/4} = rV^{1/4}(1+\epsilon)^{1/4}$ and the equation itself becomes

$$R'' + \frac{3}{16} R \left(\frac{V'}{V}\right)^2 \times K(\epsilon) = 0$$

where

$$K(\epsilon) = \frac{1 + \frac{4}{3}\epsilon + \frac{4}{3}\epsilon^2}{(1+\epsilon)^2}.$$

Chapter 3

The Determination of the Axial Potential

3.1 The functions Φ and Ψ

In order to trace the path of a particle through a lens, we have to integrate the equation of motion and it is convenient and simple to work in terms of the Picht equation. We need to know the function $T(z)$ to find the focal properties and we shall see later that we also need to know the axial derivative, $T'(z)$, if we wish to obtain information on the aberrations. The first stage in the calculation is the determination of the axial potential, from which, with its derivative, we construct $T(z)$. We shall consider ways of generating the potential distribution for lenses with more than two electrodes later, but in the first instance we concentrate our attention on lenses having only two electrodes. If the electrode potentials are V_1 and V_2, we can express the axial potential in one of the forms

$$V(z) = \tfrac{1}{2}(V_1 + V_2) + \tfrac{1}{2}(V_2 - V_1)\Phi(z)$$
$$= V_1 + (V_2 - V_1)\Psi(z)$$

where $\Phi(z)$ and $\Psi(z)$ are functions which range between ∓ 1 and between 0 and 1, respectively, as $z \to \mp\infty$.

There are various methods of calculating these functions. The general form can be approximated by a number of analytic expressions chosen more for their mathematical tractability than their connection with the basic physics of the problem. Such an approach has a long history, but with the amount of computing power readily available nowadays, there is little point in using this sort of method though we give a brief account in the next section. Results of much greater accuracy can be obtained by approximations based on the laws of electrostatics. In later sections we consider three such approaches: the solution of Laplace's equation on some discrete network of points, the analytical solution of Laplace's equation subject to boundary conditions which closely approximate the true situation, and the direct application of Coulomb's law.

20

3.2   Mathematical approximations

An exact, analytic solution of Laplace's equation exists for the case of two coaxial cylinders with negligible separation which we shall derive later. A very good approximation to the axial potential function for this case is

$$\Phi = \tanh \omega z \tag{3.1}$$

where the value of ω is found by equating $\omega \operatorname{sech}^2 \omega z$ to the slope of the true potential function at $z = 0$ and is $2.637/D$ where D is the diameter of the cylinders. Cylinders with no gap between them are somewhat impracticable, but this form for $\Phi(z)$ does give a reasonable guide to the behaviour of the lens with a small gap. Writing γ for the potential ratio V_1/V_2, the parameter, $T(z)$, of the Picht equation can be expressed as

$$T = \frac{2\omega(1 - \gamma)\exp 2\omega z}{(1 + \exp 2\omega z)(\gamma + \exp 2\omega z)} \tag{3.2}$$

which has a maximum value

$$T_m = -2\omega \left(\frac{\gamma^{1/2} - 1}{\gamma^{1/2} + 1} \right)$$

at $z_m = (1/4\omega)\ln\gamma$. Analysis using this approximate potential function usually continues by replacing equation (3.2) by one of two expressions having a similar shape:

$$T = \frac{T_m}{1 + ((z - z_m)/a)^2} \tag{3.3}$$

or

$$T = T_m \operatorname{sech}((z - z_m)/a) \tag{3.4}$$

where a is found by equating the area under the curves of equations (3.2), (3.3) and (3.4). The focal lengths and distances can then be expressed as rather complicated functions of γ, a, ω and z_m.

For the more practical case of cylinders separated by a small, but not infinitesimal, gap, g, the function

$$\Phi = \frac{1}{\omega' g} \ln \left(\frac{\cosh(\omega z + \omega' g/2)}{\cosh(\omega z - \omega' g/2)} \right) \tag{3.5}$$

has often been used. If $\omega' = \omega = 2.637/D$ this reduces to $\tanh \omega z$ as $g \to 0$. The expression for $T(z)$ is more complex and the approximation becomes progressively poorer as g increases. Some improvement may be made by modifying the value of ω' to take account of the smaller value of the slope of

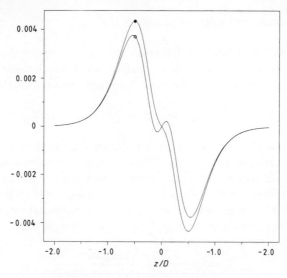

Figure 3.1   The differences between the potential functions, Φ, of equations (3.1) ($\bullet$) and (3.5) ($\circ$) and the true axial potential for a two-cylinder lens having a gap of $D/10$.

the potential at $z = 0$, and a value of $3.33/D$ has been used, but this is not really a good model for a wide gap lens. It is in any case common practice to use equations (3.3) or (3.4) for actual calculations. Figure 3.1 shows the differences between the values of Φ given by these two approximations and the true axial potential for a two-cylinder lens with a gap of $D/10$.

Some of the other approximate forms which have been suggested are:

$$\Psi = 1 - \frac{1}{\pi} \text{arccot} \left(\frac{z}{a} \right) \qquad \Psi = \left(1 + \frac{z/a}{1 + |z/a|} \right)$$

$$\Phi = \frac{1}{a} \int_0^z \exp(-\pi(z/a)^2) \mathrm{d}z$$

where the value of a depends on the length of the gap, g. There is little point in giving further discussion of this type of approximation.

3.3   The finite difference method: relaxation

One of the simplest methods for calculating the potential distribution in a bounded region is to treat the space, not as continuous, but as a lattice of discrete points. We can then replace the differential form of Laplace's equation by difference equations. We shall analyse systems of the cylindrical symmetry appropriate to electrostatic lenses later, but first consider a simpler case which allows us to illustrate some general features.

3.3.1   A two-dimensional example

Laplace's equation in rectangular cartesian coordinates is

$$\frac{\partial^2 V}{\partial x^2} + \frac{\partial^2 V}{\partial y^2} = 0. \tag{3.6}$$

We can represent this space by a rectangular array of points separated by a distance h in both the x and y directions as shown in figure 3.2, and identify five points each marked by the symbol •. The point at the centre of this group is at (x, y). We can express the potential at the four outer points using a Taylor expansion of the potential at the centre point and its derivatives at that point

$$V(x - h, y) = V(x, y) - h\frac{\partial V}{\partial x} + \frac{1}{2}h^2\frac{\partial^2 V}{\partial x^2} - \frac{1}{6}h^3\frac{\partial^3 V}{\partial x^3} + \frac{1}{24}h^4\frac{\partial^4 V}{\partial x^4} - \cdots$$

$$V(x + h, y) = V(x, y) + h\frac{\partial V}{\partial x} + \frac{1}{2}h^2\frac{\partial^2 V}{\partial x^2} + \frac{1}{6}h^3\frac{\partial^3 V}{\partial x^3} + \frac{1}{24}h^4\frac{\partial^4 V}{\partial x^4} + \cdots$$

with two similar expressions, but involving the y derivatives, for $V(x, y-h)$ and $V(x, y + h)$. Adding all four equations we find

$$V(x - h, y) + V(x + h, y) + V(x, y - h) + V(x, y + h) - 4V(x, y)$$
$$= 2h^2 \left(\frac{\partial^2 V}{\partial x^2} + \frac{\partial^2 V}{\partial y^2} \right)$$

ignoring higher-order terms beginning with

$$+\frac{1}{6}h^4 \left(\frac{\partial^4 V}{\partial x^4} + \frac{\partial^4 V}{\partial y^4} \right).$$

Equation (3.6) shows that the right hand side of this equation is zero. The potential at any point on the lattice can therefore be written as the mean of the potentials at the four nearest-neighbour points. In order to apply this result to a real problem, fixed values of the potential must be assigned to all the points on a boundary enclosing a region of our two-dimensional space. For the non-boundary points it is sometimes possible to guess approximate potentials, but unless the array is quite large it is usually enough to set all the other potentials to zero. We then move systematically over the space replacing the potential at each point by the appropriate average. It is usual to start at a place on the boundary where regions at different potentials meet. As an example, suppose that the potentials at points $(x, 10)$ and $(0, y)$ are fixed at potential 1 and those at $(0, x)$ and $(10, y)$ are fixed at zero, x and y taking all values from 1 to 9; the

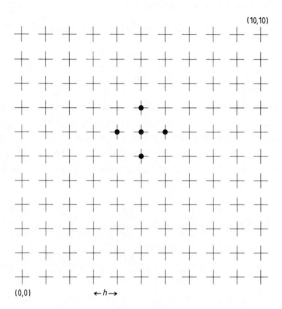

Figure 3.2　Lattice of points spaced by a distance h on which the potential may be relaxed using groups of five points ($\bullet$).

potentials at the corners are set to $0.5, 1, 0.5, 0$ in sequence, defining a square boundary. We would then begin by writing

$$V(1,1) = \tfrac{1}{4}[V(0,1) + V(2,1) + V(1,0) + V(1,2)]$$
$$= \tfrac{1}{4}[1 + 0 + 0 + 0] = 0.25$$

then increase x to the end of the row and then move up one row, return to the left and so on. A single passage across the array will not give us the solution and we must repeat the whole cycle many times until the changes from one cycle to the next are sufficiently small. This process of repeated iteration is known as relaxation. Notice the change in notation here: distances are expressed in units of h.

3.3.1.1　Improving the speed and precision.　　The relaxation method outlined above will always give a solution, but the precision will depend on the scale of the array of points, and the time taken to converge to an adequate degree will increase as the square of the number of points. This conflict between speed and precision can be resolved to some extent by first relaxing a rather coarse array and then inserting additional points between the array points, interpolating to find starting potentials for these new points, and then continuing the relaxation process. Some of the potentials we use in the programs on the accompanying disc were started with an array of

6 by 36 points and doubled four times to give a final array of 81 by 561. However, a much greater increase in speed can be obtained by a procedure known as 'successive over-relaxation'. The basis for this is that each time we calculate the potential at some point, we move it towards its final value, but by only a small amount, and indeed by an amount which is progressively smaller. Over-relaxation involves replacing the potential at a point not by the simple mean of that at its four neighbours but, recognising that next time around, it will be changed even more, we make a bigger change immediately. We write

$$V(x,y) = \frac{g}{4}[V(x-1,y) + V(x+1,y) + V(x,y-1) + V(x,y+1)]$$
$$- (g-1)V(x,y)^\dagger$$

where $V(x,y)^\dagger$ is the value of $V(x,y)$ from the preceding iteration and g is called the acceleration parameter.

As an example of the increased speed that can be obtained by an appropriate choice of acceleration parameter, we show in figure 3.3 the behaviour of a square array similar to that of figure 3.2, but with 51 points on a side. We know from symmetry that the potential of all the points on the diagonal with $x = y$ will become 0.5 eventually and so the sum of the potentials for the 49 points with $0 < x = y < 50$ will be 24.5. Figure 3.3(a) shows the exponential approach to this final value for no acceleration, and figure 3.3(b) shows, on linear scales, the sum potential for a number of values of g. With no acceleration the sum reaches a value within 0.0001 of 24.5 only after more than 3000 cycles of iteration while, for $g = 1.90$ this condition is reached after 95 cycles. Notice that the final value is reached, in this case, *from above*. The effect of increasing g from 1 to larger values resembles very much the effect of reducing the damping of a resonant circuit and, if g is increased too much, the system becomes unstable and will oscillate for $g \geq 2$; it is the first signs of this which we see for $g = 1.90$. The maximum safe value of g depends on the size of the array and also on its shape. This example has only 2401 free points and would be regarded as a rather small array. We commonly use arrays of 40 to 50 000 points with aspect ratios of 3 to 7 and have found no serious problems with values as high as 1.95. The Pascal program **RELAX51** used for this calculation is shown, in outline, on the following page, and a compiled version is included on the program disc. This version **RELAX51F.EXE** asks for a value of the acceleration parameter, g, and writes the sum potential to a file **GSXXX.POT** where **XXX** is g times 100.

```
PROGRAM relax51;

VAR
  V:array[0..50,0..50] of double;
  x,y,c:integer;
  g,Vsum:double;

PROCEDURE average;
VAR
  x,y:integer;
BEGIN
  for x:=1 to 49 do BEGIN
    for y:=1 to 49 do BEGIN
    V[x,y]:=g*(V[x+1,y]+V[x-1,y]+V[x,y+1]+V[x,y-1])/4 -(g-1)*V[x,y];
    END;
  END;
  Vsum:=0;
  for x:=1 to 49 do Vsum:=Vsum+V[x,x];
  writeln(c:4,Vsum:16:5);
END;


BEGIN
  REPEAT
    REPEAT
      clrscr;
      gotoxy(16,6);
      write('ENTER a value for the acceleration parameter, g: ');
      c:=wherex;
      writeln;
      gotoxy(16,7);
      write('(1',chr(243),' g < 2), but g=0 terminates the program');
      gotoxy(c,6);
      readln(g);
      IF g=0 THEN exit;
    UNTIL (g >= 1) and (g < 2);
    clrscr;
    c:=1;
    Vsum:=0;
    for x:=1 to 50 do for y:=0 to 49 do V[x,y]:=0;
    for x:=0 to 49 do V[x,50]:=1;
    for y:=1 to 49 do V[0,y]:=1;
    V[0,0]:=0.5;
    V[50,50]:=0.5;
    while abs(Vsum - 24.50) > 0.0001 do BEGIN
      average;
      c:=c+1;
    END;
    readln;
  UNTIL g=0;
END.
```

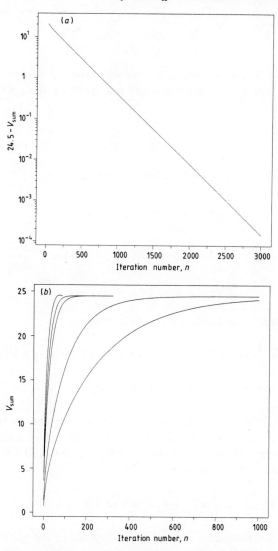

Figure 3.3   The approach of the potential on the diagonal of a square lattice (51×51 points) to the final value. (*a*) The difference between the potential after n cycles of iteration and the final potential, illustrating the exponential approach. (*b*) The potential after n cycles of iteration with acceleration parameters of (from the right) 1.0, 1.4, 1.8, 1.85, 1.90

3.3.2   Cylindrical symmetry

In cylindrical polar coordinates Laplace's equation (2.1) can be expanded as

$$\frac{\partial^2 V}{\partial z^2} + \frac{\partial^2 V}{\partial r^2} + \frac{1}{r}\frac{\partial V}{\partial r} = 0.$$

Consider a rectangular array of points separated by distances h with the polar axis forming the lower boundary and write for the potential close to the point (z, r)

$$V(z, r+h) = V(z, r) + h\frac{\partial V}{\partial r} + \frac{1}{2}h^2\frac{\partial^2 V}{\partial r^2} + \cdots.$$

Taking appropriate sums and differences we can write

$$V(z+h, r) + V(z-h, r) - 2V(z, r) = h^2\frac{\partial^2 V}{\partial z^2}$$

$$V(z, r+h) + V(z, r-h) - 2V(z, r) = h^2\frac{\partial^2 V}{\partial r^2}$$

$$V(z, r+h) - V(z, r-h) = 2h\frac{\partial V}{\partial r}$$

and substituting into Laplace's equation we have

$$V(z, r) = \frac{1}{4}[V(z+h, r) + V(z-h, r) + V(z, r+h) + V(z, r-h)]$$

$$+ \frac{h}{8r}[V(z, r+h) - V(z, r-h)]. \tag{3.7}$$

Equation (3.7) cannot be applied to points on the axis as the final term would become infinite. For this case we have to consider six points: two axial points, one to either side of the target point, and four points in a plane perpendicular to the axis.

$$V(z, 0) = \frac{1}{6}[V(z+h, 0) + V(z-h, 0) + 4V(z, h)].$$

As in the two-dimensional case, the convergence of the calculation can be made significantly faster by over-relaxation.

$$V(z, r) = \frac{g}{4}[V(z+h, r) + V(z-h, r) + V(z, r+h) + V(z, r-h)]$$

$$+ \frac{gh}{8r}[V(z, r+h) - V(z, r-h)] - (g-1)V(z, r)^\dagger \tag{3.8}$$

$$V(z, 0) = \frac{g}{6}[V(z+h, 0) + V(z-h, 0) + 4V(z, 1)] - (g-1)V(z, 0)^\dagger.$$

3.3.2.1 Boundary conditions.   It is a straightforward matter to apply the conditions at the electrode surfaces, but the gaps between electrodes can present problems and there are other 'boundaries of convenience' which can be introduced to reduce the computational space and time requirements. Considerable reductions can be made by recognising planes of symmetry in the problem. Figure 3.4 shows the electrodes of two aperture lenses, one with two apertures and the other with three. The origin of coordinates is at the centre of the lens. In the case of the two-aperture lens, the symmetry is obvious and a suitable set of boundary conditions would be $V = 0$ on the left hand electrode, $V = 0.5$ on the plane of symmetry (a boundary of convenience), and a linear rise from 0 to 0.5 along the line from a to b.

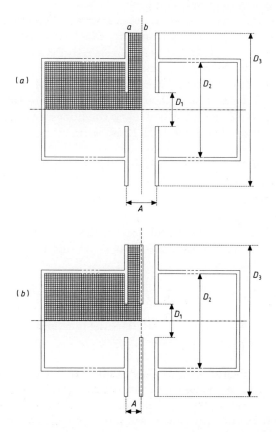

Figure 3.4 Aperture lenses of two and three elements showing the important dimensions and the regions over which the potential must be solved. The cylinders have to be bounded by a known potential at their outer ends, but the length would normally be much greater than is shown here, as is implied by the broken-line sections.

It is easy to test the validity of this last condition by applying it at a large radial distance and examining the change in potential distribution between the aperture plates as a function of radial position: the validity of the last condition is usually very good provided that $(D_3 - D_1)/A$ is greater than about 2. Only the mesh in the region indicated needs to be relaxed. The potential gradients towards the left hand end of this region are quite low and the relaxation process will converge very slowly if it is started at that end. It is far better to start the process in a region which will have large potential gradients such as the junction between the mid-plane boundary and the outer cylindrical surface as this will tend to drive the effects of the different boundaries across the mesh. One might think of the method as a type of diffusion process.

For the three-aperture lens it is necessary to find the potential distribution for two cases. The first, asymmetric, case has potentials of 0, 0.5 and 1 on the left, centre and right electrode, respectively, and a potential of 0.5 can be assigned to the centre plane leaving, again, only the hatched area to be analysed. The symmetric case, with potentials of 0, 0.5 and 0 on the three electrodes requires more thought because the potential on the centre plane is not part of the fixed boundary conditions but has to be found. This is not difficult. For $0 < r < D_1/2$ we write

$$V(0,r) = \tfrac{1}{4}[V(0,r-1) + V(0,r+1) + 2V(-1,r)]$$

and on the axis

$$V(0,0) = \tfrac{1}{2}[V(0,1) + V(-1,0)]$$

with appropriate modifications for acceleration.

If the electrode potentials are fixed at V_1, V_2 and V_3, the axial potential can then be written as

$$V(z) = V_1 + (V_3 - V_1)\Phi_a(z) + (2V_2 - V_1 - V_3)\Phi_s(z)$$

where $\Phi_a(z)$ and $\Phi_s(z)$ are the axial potential functions found for the asymmetric and symmetric cases, respectively.

3.3.2.2 Target conditions.   For the example of a square mesh considered in section 3.3.1.1, the values of the potential on all the diagonal points was known from the symmetry, and testing for convergence by examining the difference between the value after each iteration cycle and this known value was not a problem. In general the true potential distribution is not known *a priori* and the variation of the values after each cycle has to be examined. Given enough computer storage it would be possible to keep a record of all the values after a number of cycles, but this would require some multiple of the store available for calculation and is not reasonable. In the case of a lens, the important part of the potential distribution is that part on the axis and it is sufficient to look at the progressive change in these values. A good criterion is the sum of the magnitudes of the changes at all points on the axis after each cycle. There can be no false convergences as a result of cancellations and, by including all the axial points, the risk of an error due to over-rapid acceleration is low. It is in any event good practice to reduce the acceleration parameter as convergence is approached.

3.3.3   Precision and accuracy

The precision of the final result of a relaxation calculation depends on the size of the array. With other factors being constant, the precision increases directly with the number of points in the array. The *accuracy* is also affected, because the contribution of the neglected higher-order terms of the Taylor expansion becomes smaller. It is possible to increase the accuracy of the calculation by considering not just the average of the potentials at four nearest-neighbours, but by taking account of four further points. This leads to relaxation methods known as 'nine-point relaxation' in distinction to the 'five-point relaxation' we have discussed so far. If we consider four points at a distance $\pm 2h$ from the target point in the two-dimensional array [5] we can write

$$V(x-2,y) + V(x+2,y) + V(x,y-2) + V(x,y+2) - 4V(x,y)$$
$$= 8h^2 \left(\frac{\partial^2 V}{\partial x^2} + \frac{\partial^2 V}{\partial y^2} \right)$$

and this time the first neglected term is

$$+ \frac{8}{3} h^4 \left(\frac{\partial^4 V}{\partial x^4} + \frac{\partial^4 V}{\partial y^4} \right)$$

which is 16 times as great as for the four nearest-neighbour case. We can eliminate this term, leaving the sixth-order term as the first to be neglected, if we give the nearest-neighbour sum a weight of 16 and the '$2h$' sum a weight of -1

$$60V(x,y) = 16[V(x-h,y) + V(x+h,y) + V(x,y-h) + V(x,y+h)]$$
$$- [V(x-2h,y) + V(x+2h,y) + V(x,y-2h) + V(x,y+2h)].$$

This choice of additional points leads to some problems close to boundary points and it is more usual to use the points at the corners of a square to eliminate the fourth-order term. The expansion of the potential at these points to fourth order requires 15 terms:

$$V(x+h, y+h) =$$

$$V(x,y) + h\frac{\partial V}{\partial x} + h\frac{\partial V}{\partial y} + \frac{1}{2}h^2\frac{\partial^2 V}{\partial x^2} + h^2\frac{\partial^2 V}{\partial x \partial y} + \frac{1}{2}h^2\frac{\partial^2 V}{\partial y^2}$$

$$+ \frac{1}{6}h^3\frac{\partial^3 V}{\partial x^3} + \frac{1}{2}h^3\frac{\partial^3 V}{\partial x^2 \partial y} + \frac{1}{2}h^3\frac{\partial^3 V}{\partial x \partial y^2} + \frac{1}{6}h^3\frac{\partial^3 V}{\partial y^3}$$

$$+ \frac{1}{24}h^4\frac{\partial^4 V}{\partial x^4} + \frac{1}{6}h^4\frac{\partial^4 V}{\partial x^3 \partial y} + \frac{1}{4}h^4\frac{\partial^4 V}{\partial x^2 \partial y^2}$$

$$+ \frac{1}{6}h^4\frac{\partial^4 V}{\partial x \partial y^3} + \frac{1}{24}h^4\frac{\partial^4 V}{\partial y^4}.$$

We leave it as an exercise for the reader to show that the fourth-order terms can be removed by giving the nearest-neighbour terms a weight of 4 and the corner terms a weight of 1.

The nine-point versions of the finite element equations in cylindrical coordinates are quite lengthy and we shall merely quote them for the square pattern. For a general point:

$$60(8R^2 + 3)V(z,r) = 6(16R^2 + 7)[V(z+h,r) + V(z-h,r)]$$
$$+(24R^2 + 12R + 9 + 13/2R)[V(z-h,r+h) + V(z+h,r+h)]$$
$$+(24R^2 - 12R + 9 - 13/2R)[V(z-h,r-h) + V(z+h,r-h)]$$
$$+(96R^2 + 48R + 30 + 23/R)V(z,r+h)$$
$$+(96R^2 - 48R + 30 - 23/R)V(z,r-h)$$

where we have written $R = r/h$ for compactness, and for a point on the axis:

$$58V(z,0) = 34V(z,h) + 5[V(z-h,0) + V(z+h,0)]$$
$$+ 7[V(z-h,h) + V(z+h,h)].$$

The accuracy of the five-point method may be estimated by noting that the potential at a given point differs from the mean of that at the nearest-neighbours by an amount proportional to h^4 or N^{-2} where N is the number of mesh points. However, this is not the absolute accuracy since these latter points do not have their 'correct' values either so there is a progressive change which will depend on N, leaving the absolute accuracy proportional to N^{-1}. It is very easy to exploit this dependence by solving the same problem with different mesh spacing. We noted earlier that the speed of calculation could be increased by repeated halvings of the mesh separation, and if the problem is tackled in this way, the information needed to *extrapolate* to an infinite number of points is available. In figure 3.5, values of the potential at five points on the axis of a two-cylinder lens calculated with mesh spacings of $D/40$, $D/80$ and $D/160$ are plotted against the reciprocal of the number of mesh points. There is a clear dependence on N^{-1} though the slope varies and even changes sign. While this extrapolation can only be applied at the spacing of the coarsest mesh, the smooth variation of the slope can itself be interpolated so that the values of the potential at intermediate points can be corrected.

The finite difference method gives values for the potential throughout the whole space. For very many purposes it would be enough to know the axial potential, though it must be known at closely spaced points since we require the gradient in order to calculate $T(z)$. If the data are close enough, cubic spline interpolation of the potential will give a very good measure of the gradient and this is quite easy to implement. This is the reason why we favour the five-point relaxation method followed by extrapolation to infinite mesh density rather than either of the nine-point methods which might be the proper choice of finite difference method if the potential everywhere were required to high accuracy. There are particular problems close to the edges and corners of electrodes, where the potential changes rapidly. It is possible to alleviate these if the mesh density is increased locally in these regions by reducing h to a half or a quarter of its value over the main system [6]. Unfortunately, it is just at the edges of electrodes that the boundary conditions are not well known and increasing the mesh density may give a false sense of accuracy when only the *precision* of the solution to a somewhat different problem has been improved.

3.3.4   The SIMION program

SIMION is an electrostatic lens analysis and design program originally developed by D C McGilvery at LaTrobe University, Australia. It has been extensively revised† and the current version runs interactively on PC-type computers. A mouse, a maths coprocessor and an EGA display are required as are GSS CGI device drivers‡ for the display and for hard copy.

There are several parts to the program. The user can specify an electrode geometry, which may be planar or have cylindrical symmetry, and assign potentials. The program uses five-point relaxation on a mesh having at most 16 000 points. The user can specify an acceleration parameter and the program will adjust this dynamically, increasing it in the early stages and decreasing it as convergence is approached. The test for convergence is that the change in potential at no point should exceed some user specified

---

† Information about SIMION and copies of the program may be obtained from David A Dahl, MS2208, Idaho National Engineering Laboratory, EG&G Idaho Inc., Idaho Falls, ID 83415, USA.

‡ Graphics Software Systems, 9590 SW Gemini Drive, P.O. Box 4900, Beaverton, Oregon 97005-7161, USA.

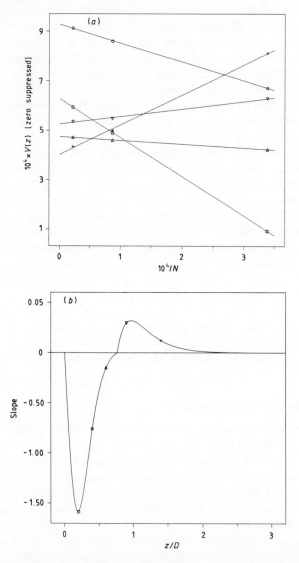

Figure 3.5 (*a*) The potential at five points on the axis of a cylindrical lens calculated by five-point relaxation on meshes having spacings of *D*/40, *D*/80, and *D*/160 plotted against the reciprocal of the number of mesh points, illustrating extrapolation to an infinite mesh density. The five sets of data are shown with substantially suppressed zeros to emphasise the variation in slope. The data for 0.9 and 1.4 diameters are shown on ordinate scales magnified by 10 and 100, respectively. (*b*) The dependence of the slope of the lines of (*a*) (identified by their symbols) and similar plots on axial position. The scale is magnified by 10 for the positive values.

value. After convergence, equipotential lines can be drawn and trajectories of ions or electrons plotted. The trajectories are drawn in real space and use the potential along the paths and not merely the axial potential. This means that spherical aberration can be shown very clearly by launching rays at various angles to the axis. There is a brief description of the use of **SIMION** for this type of study in Chapter 5. Multi-element lenses can be handled by solving the potential problem for electrode potentials of 0, 1, 2... and are then adjusted very rapidly by superposition when realistic potentials are used. Ray tracing is done using a fourth-order Runge–Kutta method, using forces calculated from the potentials at the four nearest points as the trajectory evolves. Some of the figures in this book have been prepared using **SIMION** and output to a 300 dot per inch printer. Multicolour pictures may be produced on a graph plotter with the appropriate driver.

3.4 Bessel function expansions

The potential, $V(z, r)$, can be written as a constant term plus the product of separate functions, $Z(z)$ and $R(r)$, of the axial and radial coordinates. If we differentiate this product and substitute into Laplace's equation we obtain an equation in which the terms in z and in r may each be set equal to some parameter which we write as $-k^2$

$$\frac{1}{R}\left(\frac{\mathrm{d}^2 R}{\mathrm{d}r^2} + \frac{1}{r}\frac{\mathrm{d}R}{\mathrm{d}r}\right) = -\frac{1}{Z}\frac{\mathrm{d}^2 Z}{\mathrm{d}z^2} = -k^2.$$

The differential equations for the two functions are then

$$\frac{\mathrm{d}^2 Z}{\mathrm{d}z^2} - k^2 Z = 0 \qquad \frac{\mathrm{d}^2 R}{\mathrm{d}(kr)^2} + \frac{1}{kr}\frac{\mathrm{d}R}{\mathrm{d}(kr)} + R = 0.$$

The solution for $Z(z)$ is straightforward

$$Z(z) = A_k \mathrm{e}^{kz} + B_k \mathrm{e}^{-kz}$$

while that for $R(r)$ requires some consideration of the context. The equation is Bessel's equation of zero order for which the formal solution would be

$$R(kr) = C_1 J_0(kr) + C_2 N_0(kr)$$

where J_0 and N_0 are Bessel functions of zero order and of the first and second kind, respectively. The latter becomes infinite for zero argument and can not therefore give a physically real description of the potential in a charge-free region, so $C_2 = 0$ and the overall solution for the potential becomes

$$V(z,r) = (A_k e^{kz} + B_k e^{-kz}) J_0(kr) \tag{3.9}$$

where we have incorporated the coefficient C_1 into A_k and B_k. Notice that the zero-order Bessel function of the first kind can be expressed as

$$J_0(r) = \sum_{i=0}^{\infty} \frac{(-1)^i (r/2)^{2i}}{(i!)^2} = 1 - \frac{r^2}{2} + \frac{r^4}{64} - \frac{r^6}{2304} + \cdots$$

which is exactly the form given in equation (2.2) by the Taylor expansion. The general solution for a particular lens geometry will consist of a sum of such terms with, perhaps, a constant representing the potential of an electrode. Figure 3.6 shows the geometry of a lens formed by two coaxial cylinders of equal diameter, D, separated by an axial distance, g, and having potentials V_1 and V_2. We show the walls of the cylinder to be of finite thickness and shall explain the significance of this later. Three regions are indicated, in each of which we can write a specific expression for the potential. In regions I and III, within the cylinders, only one of the terms of equation (3.9) has physical meaning; the other would become infinite as $|z| \to \infty$.

$$V_{\mathrm{I}}(z,r) = V_1 + \sum_{n=1}^{\infty} A_n \exp(k_n z) J_0(k_n r) \tag{3.10a}$$

$$V_{\mathrm{II}}(z,r) = \frac{V_1 + V_2}{2} + \left(\frac{V_2 - V_1}{g} \right) z$$

$$+ \sum_{n=1}^{\infty} [B_n \exp(-k_n z) + B_n' \exp(k_n z)] J_0(k_n r) \tag{3.10b}$$

$$V_{\mathrm{III}}(z,r) = V_2 + \sum_{n=1}^{\infty} C_n \exp(-k_n z) J_0(k_n r). \tag{3.10c}$$

In order to find the coefficients, we must consider the conditions at the various boundaries. At the inner surfaces of the cylinders the potentials are V_1 and V_2 so, unless $A_n = C_n = 0$, the Bessel functions must all vanish for $r = D/2$. The possible values of $k_n D/2$ are therefore the roots, $j_{0,n}$, of $J_0 = 0$ and will be found in tables of Bessel functions. This restriction on the values of k_n means that in region II the potential at $r = D/2$ varies linearly with z. This is not correct, but it is not a bad approximation for small values of g/D and allowances can be made. Symmetry allows us to write

$$A_n = -C_n \quad \text{and} \quad B_n = -B_n'$$

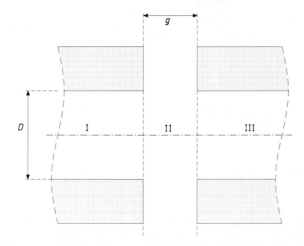

Figure 3.6   Schematic diagram of a two-cylinder lens showing the dimensions and identifying the three regions to which different expressions for the potential apply. The origin of the coordinates is at the centre of the system.

and from the continuity of potential at the boundary between regions I and II we have

$$A_n \exp(-k_n g/2) = B_n[\exp(k_n g/2) - \exp(-k_n g/2)]$$

which allows us to express the potentials in all three regions in terms of the single set of coefficients B_n

$$V_{\text{I}}(z, r) = V_1 + \sum_{n=1}^{\infty} B_n[\exp(k_n g) - 1] \exp(k_n z) J_0(k_n r)$$

$$V_{\text{II}}(z, r) = \frac{V_1 + V_2}{2} + \left(\frac{V_2 - V_1}{g}\right) z$$

$$+ \sum_{n=1}^{\infty} B_n[\exp(-k_n z) - \exp(k_n z)] J_0(k_n r)$$

$$V_{\text{III}}(z, r) = V_2 - \sum_{n=1}^{\infty} B_n[\exp(k_n g) - 1] \exp(-k_n z) J_0(k_n r).$$

The continuity of potential across the boundary between regions I and II implies that the radial component of the field is continuous across this boundary, but we have not used the constraint that the axial component of the field must also be continuous across this boundary,

$$\sum_{n=1}^{\infty} B_n k_n \exp(k_n g/2) J_0(k_n r) = \frac{V_2 - V_1}{2g} \qquad \text{for all } r \le D/2.$$

It is not easy to apply this constraint directly, but another approach allows all the B_n to be determined.

3.4.1   The variational principle

It is a general feature of any physical system that it will adjust itself to a condition of minimum potential energy. We may therefore test the suitability of any expression for the potential by examining the energy stored in the associated field. The energy density of an electrostatic field can be written as $\frac{1}{2}\epsilon_0 E^2$. If we write the total energy of the field as the integral of the sums of the squares of the axial and radial differentials of the potentials we can determine the coefficients by equating the differentials of this integral (with respect to B_n) to zero. The expression for the energy, W, contains three terms, one for each of the regions, but we need only evaluate one and a half of these because of the symmetry about the central plane.

$$W = \pi\epsilon_0 \left\{ \int_{-\infty}^{-g/2} \int_{0}^{D/2} \left[\left(\frac{\partial V_{\mathrm{I}}}{\partial z}\right)^2 + \left(\frac{\partial V_{\mathrm{I}}}{\partial r}\right)^2 \right] r\,dr\,dz \right.$$

$$\left. + \int_{-g/2}^{0} \int_{0}^{D/2} \left[\left(\frac{\partial V_{\mathrm{II}}}{\partial z}\right)^2 + \left(\frac{\partial V_{\mathrm{II}}}{\partial r}\right)^2 \right] r\,dr\,dz \right\}.$$

The evaluation of these integrals is considerably simplified by the orthogonality of the Bessel functions.

Applying the condition $\partial W/\partial B_n = 0$ gives

$$B_n = (V_2 - V_1)\frac{\exp(-k_n g/2)}{(g/2)k_n^2 D J_1(k_n D/2)}$$

where $J_1(k_n D/2)$ is the Bessel function of order one evaluated at the zeros of J_0.

The properties of the field do not depend on the scale of the system and it is convenient to express all the dimensions in terms of the diameter, D, of the cylinders

$$G = g/D \qquad Z = z/D \qquad R = 2r/D$$

with

$$K_n = k_n D/2.$$

Values of K_n and the corresponding values of $J_1(K_n)$ to ten decimal places can be found in tables [7] for $1 \leq n \leq 150$. Writing $Q_n = \frac{1}{2}K_n^2 J_1(K_n)$

we have $B_n = (V_2 - V_1)Q_n \exp(-K_n G)/G$. The exponential terms involve $(\pm 2Z \pm G)$ so we write

$$a1_n = \exp[K_n(2Z + G)] \qquad \text{and} \qquad a2_n = \exp[K_n(2Z - G)]$$

and express the potentials as

$$V_I(Z, R) = V_1 + \frac{V_2 - V_1}{G} \sum_{n=1}^{\infty} Q_n J_0(K_n R)(a1_n - a2_n) \qquad (3.11a)$$

$$V_{II}(Z, R) = \frac{V_1 + V_2}{2} + \frac{V_2 - V_1}{G} Z$$
$$+ \frac{V_2 - V_1}{G} \sum_{n=1}^{\infty} Q_n J_0(K_n R)(1/a1_n - a2_n) \qquad (3.11b)$$

$$V_{III}(Z, R) = V_2 - \frac{V_2 - V_1}{G} \sum_{n=1}^{\infty} Q_n J_0(K_n R)(1/a2_n - 1/a1_n). \ (3.11c)$$

Successive terms of the summations converge quite rapidly except close to the boundaries between the regions, and at axial distances greater than some $1\frac{1}{2}$ diameters only the first term makes any significant contribution. The axial potential then approaches that of the electrode as $e^{-2K_0 Z}$ where $K_0 = 2.4048$. As $G \to 0$ the exponential terms approach $2K_n G \exp(-2K_n Z)$, the $(2K_n G)$ cancels, and the expressions reduce to

$$2 \sum_{n=1}^{\infty} \frac{\exp(-2K_n|Z|)}{K_n J_1(K_n)} J_0(K_n R).$$

3.4.1.1 The problem of the non-linear potential in the gap. For small gaps the departure of the true potential between the cylinders at $r = D/2$ from a linear dependence on z is not really serious, but nonetheless it is possible to make a correction in one of two ways. Instead of writing the second term of equation (3.10b) as $((V_2 - V_1)/g)z$, a third-order term can be added [8], and the variational calculation repeated with the slightly more complicated expression

$$(V_2 - V_1) \left[\left(1 - \frac{a}{2}\right) \frac{z}{g} + 2a \left(\frac{z}{g}\right)^3 \right].$$

For a zero gap lens $a = 1$, and decreases quite slowly, being about 0.75 for $g = D/2$. This is too simple to represent the true situation except, perhaps, for cylinders with extremely thin walls. Thick-walled cylinders are more practicable and have the further virtues of shielding the cylindrical space from charged surfaces outside the lens system, but within the vacuum chamber. The true variation of the potential at $r = D/2$ may be found by relaxing the potential not only within the cylindrical surface, but outwards to a greater radius within the gap as discussed in section 3.3.2.1 in the context of aperture lenses. Figure 3.7 illustrates this variation for a gap, $g = D/10$. The cubic form is clearly a poorer representation than the simple linear variation and the true shape requires higher-order terms to describe it. It would be quite difficult to repeat the variational calculation with fifth- and seventh-order terms but the straight line of slope 9.238 provides a guide to improving the approximation. Instead of using the true value of the gap in the expressions for the potential, we use a value of $1.0825 \times g$. This gives an axial potential which agrees with that calculated by five-point relaxation with extrapolation, to 1 or 2 parts in the sixth decimal place for electrode potentials of 0 and 1.

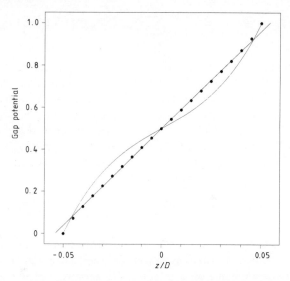

Figure 3.7   The potential in the gap between two cylinders at potentials 0 and 1. The full curve shows the 'cubic' result of Bonjour [8] and the points show our relaxation results. A straight line of slope 9.238 is also shown: its significance is discussed in the text.

3.5   The charge density method

The solution of Laplace's equation requires the conditions to be known over a boundary which completely encloses the space. Almost without exception this calls for some approximation of the conditions in the gaps between the electrodes. The charge density method does not have this constraint. It is in principle a very simple method: any electrode may be replaced by a system of charges at the electrode surface provided that the potentials they produce are the same everywhere as those produced by the electrode. In particular, the potential of the electrode itself must be reproduced by the charge distribution. The formal way to express this is

$$V_k = \frac{1}{4\pi\epsilon_0} \sum_{j=1}^{n} \int_{S_j} \frac{\sigma_j(\mathbf{r}_j)\mathrm{d}S_j}{|\mathbf{r}_j - \mathbf{r}_k|} \tag{3.12}$$

where V_k is the potential at a position $\mathbf{r}_k$ due to charges $q_j = \sigma_j \mathrm{d}S_j$ at positions $\mathbf{r}_j$ on each of n electrodes. In the present context of round lenses with axial symmetry, the problem reduces to the summation of the potentials at a point due to rings of charge on all the electrodes. The geometry of this situation is illustrated in figure 3.8 in which the potential at the ring of radius r_i due to the charge on the ring of radius r_j is to be found. In terms of the distances shown in this diagram the denominator of equation (3.12) is

$$l = [r_i^2 + r_j^2 - 2r_i r_j \cos(\alpha_i - \alpha_j) + (z_i - z_j)^2]^{1/2}.$$

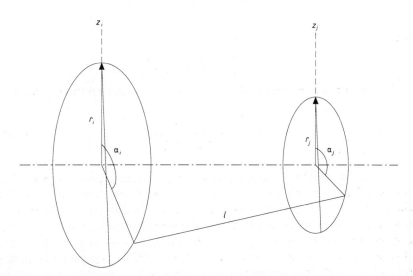

Figure 3.8   Diagram to indicate the coordinates appropriate to the calculation of the potential at a point due to a ring of charge.

The absolute values of the angles are arbitrary because of the axial symmetry so, for convenience, let $\alpha_i = \pi$ which allows us to simplify this expression as

$$l = [(r_i + r_j)^2 - 4r_i r_j \sin^2(\alpha_j/2) + (z_i - z_j)^2]^{1/2}$$

$$= \left[[(r_i + r_j)^2 + (z_i - z_j)^2] \left(1 - \frac{4r_i r_j \sin^2(\alpha_j/2)}{(r_i + r_j)^2 + (z_i - z_j)^2} \right) \right]^{1/2}.$$

The total charge on the ring, Q_j, can be written in terms of a line density, $\lambda_j = Q_j/2\pi r_j$, and each term in the summation of equation (3.12) becomes

$$\frac{1}{4\pi\epsilon_0} \int_0^{2\pi} \frac{\lambda_j r_j \, d\alpha_j}{l}$$

$$= \frac{1}{\pi\epsilon_0} \frac{\lambda_j r_j}{[(r_i + r_j)^2 + (z_i - z_j)^2]^{1/2}}$$

$$\times \int_0^{\pi/2} \left(1 - \frac{4r_i r_j \sin^2(\alpha_j/2)}{(r_i + r_j)^2 + (z_i + z_j)^2} \right)^{-1/2} d(\alpha_j/2).$$

The integral is just $K(\tau)$, the complete elliptic integral of the first kind,

$$\int_0^{\pi/2} (1 - \tau^2 \sin^2 \beta)^{-1/2} d\beta$$

with

$$\tau^2 = \frac{4r_i r_j}{(r_i + r_j)^2 + (z_i - z_j)^2}$$

and so in terms of τ and Q_j the potential at a point i can be written as

$$V_i = \sum_{j=1}^{n} \frac{Q_j}{4\pi^2 \epsilon_0} \frac{\tau}{\sqrt{r_i r_j}} K(\tau). \tag{3.13}$$

The practicability of solving this set of equations depends on a sensible choice of 'ring' elements and the distribution of charge between them. The charge density on the surface of an electrode is directly proportional to the electric field at each point and a preliminary estimate of this field may be made in other ways, such as by the finite difference method. The charge density near the edge of an electrode will be high and will change rapidly so small rings are needed here: conversely, deep inside a cylindrical electrode the charge density will be small and a substantial length of the cylinder may be used as a single 'ring'. This choice of elements is probably the factor which most severely limits the overall accuracy of the method. The most direct way of solving the problem is to treat equation (3.13) as a matrix equation relating the column vectors $\boldsymbol{V}_i$ and $\boldsymbol{Q}_j$, with the individual elements being functions of the positions of the two rings involved, but other methods have been used [9].

Once the charge distribution consistent with the required electrode potentials has been found, the axial potential distribution follows readily, since the elliptic integrals are then just equal to 1. The gradient of the axial potential can also be expressed in closed form and the function $T(z)$ of the Picht equation follows immediately.

Chapter 4

The Optics of Simple Lens Systems

4.1 Aperture lenses

One might almost say that aperture lenses are the 'traditional' particle lens. A great variety of detailed shapes and proportions have been used which makes it difficult to prepare tables of focal data for general application. In the context of low energy beams there is no need for smoothly rounded electrodes: simple thin discs with circular holes can be used and, provided that there is some agreement about the spacing of the discs, it should be possible to make systems of two or three apertures which behave in a predictable fashion. Figure 3.4 shows the important dimensions of such systems. For the purposes of scaling, it is usual to express all the dimensions in terms of the diameter, D_1, of the apertures themselves. The significance of the outer diameter of the discs, D_3, was discussed earlier in the context of establishing a simple boundary condition to use in the relaxation calculation, but it has a further, practical, significance in that the lens system does not exist in a vacuum, but rather in a vacuum *system* which has walls which will either be of metal and therefore have some definite potential, or of an insulating material and have a rather indefinite potential. In both cases it is important that this should not influence the potential distribution within the lens and it is as well to ensure that $(D_3 - D_1)/A > 3$.

A similar argument requires that the first and last apertures be mounted on cylinders to guarantee uniform potentials in the object and image spaces, and the diameter, D_2, of these cylinders affects the potential distribution and, therefore to some extent the focusing properties. Figure 4.1 shows the axial potential distribution in a two-aperture lens with $A = D_1/2$ and a number of values of D_2/D_1. Near the centre of the lens the distributions are very similar, but with increasing distance from the centre, the potential tends towards the simple exponential decay characteristic of the dominant term in the Bessel function expansion, $V(z, r) \propto \exp(-k_0 z / \frac{1}{2} D_2)$. Most of the calculations were made with the length of the D_2-cylinder great enough that the potential fell smoothly, but one example is included of a cylinder

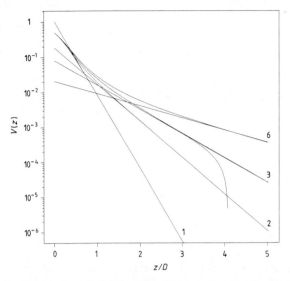

Figure 4.1 Axial potential distributions for two-aperture lenses having $A = D_1/2$. Values of the ratio D_2/D_1 are shown on the diagram. The potentials all have the value 0.5 at the lens centre, but fall with increasing axial distance, z, as $\exp(-4.8087(z/D_2))$. The potentials approach these asymptotes from above, except in the case of the cylinder lens.

of length $4D_1$, and therefore only $4D_2/3$ long, to illustrate the way in which the end plate can influence the potential distribution.

The different potential distributions away from the apertures themselves affect the properties of the lens as illustrated in figure 4.2. This shows the first focal length, f_1, the mid-focal distance, F_1, and the spherical aberration coefficient, C_{s_0}, for a voltage ratio, $V_2/V_1 = 10$, as functions of the ratio D_2/D_1, and includes the case of the cylinder lens of the same spacing, which one might consider to be a rather extreme aperture lens having $D_2/D_1 = 1$. The overall effect on the focal properties is not large, but the lower aberration of the cylinder lens is quite noticeable. Changes of a similar scale result if the separation of the apertures is doubled, but whereas this is a fairly obvious effect, the effect of changes in D_2 is a little more subtle. The thickness of the aperture discs also plays a part, but not on the same scale. The data presented in the **LENSYS** program are for $D_2/D_1 = 3$ which represents a compromise between reduced dependence at large values of the ratio and sheer physical size of the electrodes.

4.1.1   Three-aperture lenses

As is the case with all lenses with three elements, the principal function is the decoupling of the focal properties of the lens, in particular the behaviour

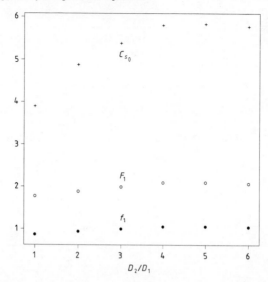

Figure 4.2   Focal lengths, (o), mid-focal distances, ($\bullet$), and spherical aberration coefficients, ($+$), for two-aperture lenses having $A = D_1/2$ and $V_2/V_1 = 10$.

between conjugates, from the ratio of the potentials of the image and object spaces. Data for this sort of application are generally presented in the form of plots of the ratio V_2/V_1 against V_3/V_1. These plots are usually called 'focal loci' or 'zoom curves' and lenses used in this fashion are referred to as 'zoom lenses', but the behaviour is more restricted than the use of this term in light optics would imply. We shall return to this topic in section 4.6.1.

4.1.1.1   The einzel lens.   Nowadays this term is understood to refer to a three-element lens operated symmetrically, with $V_3 = V_1$ and with V_2 taking any appropriate value. Historically the einzel, or unipotential, lens signified a system with V_2 held at the potential of the cathode. The major advantage of this is that the focal properties, depending as they do on voltage *ratios*, are the same for all values of $V_1 = V_3$. The disadvantage is that the aberrations of this lens are greater than those of the symmetric lens with $V_2 > V_{1,3}$.

The potential distribution in an einzel lens is different in character from that shown for the two-cylinder lens in figure 1.2(b) and we illustrate this in figure 4.3. Apart from the bowing out of the equipotentials which occurs for any aperture lens, the potential close to the centre forms a saddle where the potential rises (say) in the r direction and falls in the $\pm z$ direction. This general form will apply to any three-aperture lens for which V_2 is outside the range spanned by V_1 and V_3. One of the equipotentials crosses the axis at the centre and it is simple to show that the angle between this equipotential (the value of which will depend on the thickness of the aperture plates) and the axis is ±54.7°. Expanding the axial potential as

$$V(z,0) = V_0(z,0) + zV_0'(z,0) + \tfrac{1}{2}z^2V_0''(z,0) + \cdots$$

and substituting this into equation (2.2) gives

$$V(z,r) = V_0(z,0) + zV_0'(z,0) + \tfrac{1}{2}z^2V_0''(z,0) - \tfrac{1}{4}r^2V_0''(z,0) + \cdots.$$

$V_0'(z,0) = 0$ at the saddle point and $V(z,r) = V_0(z,0)$ along the equipotential, so the equation of the equipotential close to the saddle point is $\tfrac{1}{2}z^2 = \tfrac{1}{4}r^2$ from which the result for the angle follows immediately.

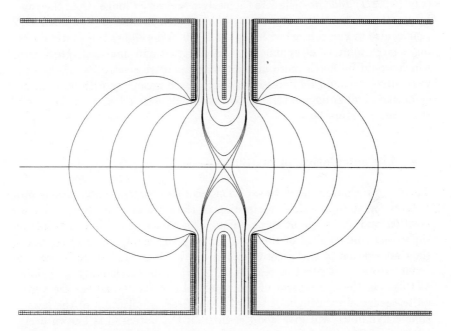

Figure 4.3   Equipotentials in an einzel lens. Near the centre of the lens the equipotentials are at 0.300, 0.294 and 0.288 of the potential difference between the outer and centre electrodes. The extreme equipotentials are at 0.020 of this difference.

4.2   Cylinder lenses

Cylinder lenses offer a number of advantages over aperture lenses of the same inner diameter. For a given voltage ratio they are a little stronger and have smaller aberration coefficients, but there are two quite important mechanical advantages as well. They are easy to make and to mount: it is easier to turn cylinders on a lathe than to punch clean-edged holes in metal sheet. It is also simple to make lens systems with rather small inter-electrode gaps. This is an important factor in keeping the potential distribution in the gap at the inner surface of the electrodes close to that assumed in the calculation of the entire potential distribution and also ensures good screening from any stray electric fields around the lens structure.

The potential and field distributions in a system of several coaxial cylinders can be found quite adequately by the superposition of those for contiguous pairs, provided that the gaps between the cylinders are small (say, $\not> 0.2D$) and the cylinders themselves are rather longer than the gaps. In practice it is common to use cylindrical elements no shorter than $D/2$ (gap centre to gap centre) with gaps of $D/10$. This allows lens systems having a large number of cylinders to be designed and analysed, procedures which would be much more difficult with aperture electrodes. If the gaps were large, it would be necessary to solve each geometry with appropriate symmetric and antisymmetric potentials on all electrodes, as in the case of the three-aperture lens.

4.3   The calculation of ray paths

There are no purely analytic solutions to this problem, only various procedures for the numerical integration of the path starting from a known position and the slope at some point. The choice of method depends on the extent and quality of the data. If only the axial potential is known, then integration of the Picht equation can be done, and there is one procedure (to be described in section 4.3.2) which is particularly well suited to this. On the other hand, if the potential and its derivatives are known quite accurately throughout the space, a direct application of the laws of simple dynamics will give good results. If the potential is known everywhere, but to limited accuracy, various standard procedures are available. However, care should be taken to distinguish between the precision of the integration procedure and the accuracy of the trajectory. Ray tracing in the *correct* potential will give results which show the effects of aberrations, while integration of the Picht equation will give only the paraxial, Gaussian behaviour, though corrections can be applied and aberration coefficients calculated even in this case.

4.3.1   Ray tracing in a known potential

If the potential is not known really well throughout the space, the best approach is to use some standard expression for the integration of the differential equations, but if it and its derivatives are well known an analysis by direct application of the laws of dynamics using expansions of the coordinates and velocities about each point will give good results [10].

4.3.1.1 Integration of the equation of motion.   Taking the time, t, to be the independent variable we write the coordinates as

$$z(t) = z_0 + t\left(\frac{dz}{dt}\right)_0 + \frac{t^2}{2}\left(\frac{d^2z}{dt^2}\right)_0 + \frac{t^3}{6}\left(\frac{d^3z}{dt^3}\right)_0 + \cdots \quad (4.1)$$

with a similar expression for $r(t)$, where $z(t)$ and $r(t)$ are the axial and radial positions of a particle (having initial coordinates denoted by the '0' subscripts) after a short time interval t. The components of the velocity follow from differentiation of these equations

$$\frac{dz(t)}{dt} = \left(\frac{dz}{dt}\right)_0 + t\left(\frac{d^2z}{dt^2}\right)_0 + \frac{t^2}{2}\left(\frac{d^3z}{dt^3}\right)_0 . \quad (4.2)$$

The accelerations are given by

$$\left(\frac{d^2z}{dt^2}\right)_0 = \frac{e}{m}\frac{\partial V(z,r)}{\partial z}$$

and the third time derivatives of the coordinates are therefore

$$\left(\frac{d^3z}{dt^3}\right)_0 = \frac{e}{m}\left(\frac{\partial^2 V}{\partial z^2}\right)_0\left(\frac{dz}{dt}\right)_0 + \frac{e}{m}\left(\frac{\partial^2 V}{\partial z\partial r}\right)_0\left(\frac{dz}{dt}\right)_0 .$$

Substituting these expressions into equations (4.1) and (4.2), and introducing a 'reduced time' $\tau = t\sqrt{2e/m}$ to avoid the e/m which occurs in nearly every term, we construct the following recurrence relationships

$$z(t) = z_0 + \tau\left(\frac{dz}{d\tau}\right)_0 + \frac{\tau^2}{4}\left(\frac{\partial V}{\partial z}\right)_0$$
$$+ \frac{\tau^3}{12}\left[\left(\frac{\partial^2 V}{\partial z^2}\right)_0\left(\frac{dz}{d\tau}\right)_0 + \left(\frac{\partial^2 V}{\partial z\partial r}\right)_0\left(\frac{dr}{d\tau}\right)_0\right]$$

$$\frac{dz(t)}{d\tau} = \left(\frac{dz}{d\tau}\right)_0 + \frac{\tau}{2}\left(\frac{\partial V}{\partial z}\right)_0$$
$$+ \frac{\tau^2}{4}\left[\left(\frac{\partial^2 V}{\partial z^2}\right)_0\left(\frac{dz}{d\tau}\right)_0 + \left(\frac{\partial^2 V}{\partial z\partial r}\right)_0\left(\frac{dr}{d\tau}\right)_0\right]$$

with two similar expressions for the radial coordinate and velocity.

Only expressions for the potential which can be evaluated and differentiated analytically at any point such as those expressed in terms of Bessel functions or elliptic integrals are likely to be good enough to yield accurate values for the second derivative required for this type of analysis. The calculations are surprisingly simple and we illustrate this using the potential in the gap region of a two-cylinder lens developed in section 3.4.1 (equation (3.11b))

$$V_{\text{II}}(Z, R) = \frac{V_1 + V_2}{2} + \frac{V_2 - V_1}{G} Z + \frac{V_2 - V_1}{G} \sum_{n=1}^{\infty} Q_n J_0(K_n R)(1/a1_n - a2_n).$$

The five differential expressions can be written as

$$\frac{\partial V}{\partial Z} = \frac{(V_2 - V_1)}{G} - \frac{2(V_2 - V_1)}{G} \sum_{n=1}^{\infty} Q_n K_n J_0(K_n R)(1/a1_n + a2_n)$$

$$\frac{\partial V}{\partial R} = -\frac{2(V_2 - V_1)}{G} \sum_{n=1}^{\infty} Q_n K_n J_1(K_n R)(1/a1_n - a2_n)$$

$$\frac{\partial^2 V}{\partial Z^2} = \frac{4(V_2 - V_1)}{G} \sum_{n=1}^{\infty} Q_n K_n^2 J_0(K_n R)(1/a1_n - a2_n)$$

$$\frac{\partial^2 V}{\partial R^2} = -\frac{2}{R} \frac{\partial V}{\partial R} - \frac{\partial^2 V}{\partial z^2}$$

$$\frac{\partial^2 V}{\partial Z \partial R} = \frac{4(V_2 - V_1)}{G} \sum_{n=1}^{\infty} Q_n K_n^2 J_1(K_n R)(1/a1_n - a2_n).$$

Very similar expressions apply in the outer regions, the principal differences arising from the different exponents.

The accuracy of the ray integration depends on the interval in reduced time, $\Delta\tau$, which is related to the step length, Δl, by $\Delta\tau = \Delta l/V^{1/2}$ and it is important to choose a suitable value. Too short an interval will waste computer time, but the optimum interval will depend on the strength of the field and so it needs to be adjusted dynamically. The accuracy of the calculations is open to a simple test: the difference in potential between the initial and final points of each step should be equal to the difference in the squares of the velocities of the particle multiplied by $m/2e$. If the fractional difference found after a step is greater than some chosen value, then the step should be recalculated with $\Delta\tau$ halved. If the difference exceeds some smaller criterion, $\Delta\tau$ may be doubled for the next step.

4.3.1.2 Runge–Kutta methods.

Unless we have a good enough potential we can not make proper use of terms beyond the first order of the Taylor expansion of equation (4.1) and need to find some other way to make allowance for the higher-order terms. The Runge–Kutta methods are based on the replacement of these terms by first-order expansions at nearby points. Figure 4.4 illustrates the steps required to apply these methods. The function describing some aspect of the trajectory is represented by a curve starting at the point A, (x_n, y_n), and we wish to find the y coordinate at C, for which $x_{n+h} = x_n + h$. We can write this as $y_{n+1} = y_n + k$ where k is h times some mean gradient. The very first approximation would be to write $k = k_1 = hy'(x_n, y_n)$ where the gradient of the function is evaluated at the point $A(x_n, y_n)$. This would take us to the point D_1 and is plainly not a good approximation.

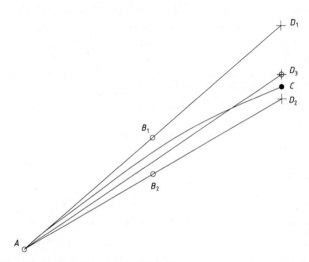

Figure 4.4   A schematic illustration of the steps in the fourth-order Runge–Kutta method. AC represents the real trajectory and AD_1 is the tangent to the trajectory at A. The gradient of the function at B_1 is used to find D_2, the second approximation to C, then the gradients at B_2 and D_3 are evaluated and a weighted sum of the four gradients is then a good approximation to that of the chord AC.

However, common sense and the mean value theorem tell us that *at some point* along the curve the tangent is parallel to the chord AC joining the initial and the true final points. The next stage of approximation is therefore to evaluate k at the mid-point, B_1, of the chord AD_1 and write $k = k_2 = hy'(x_n + h/2, y_n + k_1/2)$ which will take us to the point D_2. This is a significantly better approximation and the errors are of order h^3: it is therefore referred to as the second-order Runge–Kutta method. It is worth making two further steps of this sort since the gain in precision outweighs the complexity of the calculation, but beyond that point one should really question whether the potential function is good enough to justify the effort.

The four successive values for k can be summarised as

$$
\begin{aligned}
k_1 &= hy'(x_n, y_n) & \text{at } A \\
k_2 &= hy'(x_n + h/2, y_n + k_1/2) & \text{at } B_1 \\
k_3 &= hy'(x_n + h/2, y_n + k_2/2) & \text{at } B_2 \\
k_4 &= hy'(x_n + h, y_n + k_3) & \text{at } D_3.
\end{aligned}
$$

These expressions have a nice symmetry and we write

$$
k = \frac{k_1}{6} + \frac{k_2}{3} + \frac{k_3}{3} + \frac{k_4}{6}
$$

where the weights given to the k_i have been chosen such that the third- and fourth-order terms in the Taylor expansions cancel and the errors are now of order h^5 giving the fourth-order Runge–Kutta method.

To apply this to the tracing of ray paths, we write the axial, z_n, and radial, r_n, positions and the corresponding velocities, u_n and v_n, respectively, in terms of the reduced time, τ, as before (i.e. $u_n = dz/d\tau$, etc) and apply the recurrence relations to all four variables.

$$
z_{n+1} = z_n + j \qquad r_{n+1} = r_n + k \qquad u_{n+1} = u_n + l \qquad v_{n+1} = v_n + m
$$

where j, k, l and m are the appropriate weighted sums. The individual approximations are

$$
\begin{aligned}
j_1 &= hu_n & l_1 &= -hE_z(z_n, r_n)/2 \\
j_2 &= h(u_n + l_1/2) & l_2 &= -hE_z(z_n + l_1/2, r_n + j_1/2)/2 \\
j_3 &= h(u_n + l_2/2) & l_3 &= -hE_z(z_n + l_2/2, r_n + j_2/2)/2 \\
j_4 &= h(u_n + l_3) & l_4 &= -hE_z(z_n + l_3, r_n + j_3)/2
\end{aligned}
$$

with a similar set for the radial motion. h is an interval of reduced time, and E_z and E_r are the axial and radial components of the electric field strength which must be found at all the indicated points by numerical differentiation of the potential.

4.3.2 Integration of the Picht equation

The Picht equation describes the evolution of a reduced radius defined by $R(z) = r(z)V^{1/4}(z,0)$ as a function of z. It involves the axial potential, $V(z,0)$ and its first derivative and contains only the second derivative of $R(z)$. There is a standard procedure for the integration of such equations which is faster than fourth-order Runge–Kutta with the same step size and is effectively of higher order. We write the Picht equation as

$$
\frac{d^2 R(z)}{dz^2} = -T^*(z)R(z)
$$

where

$$T^*(z) = \frac{3}{16}\left(\frac{V'(z,0)}{V(z,0)}\right)^2.$$

$T^*(z) = \frac{3}{16}T(z)^2$, where $T(z)$ was introduced in Chapter 2. Some authors use $T(z)$ where we use $T^*(z)$, but it is necessary to distinguish the two symbols since $T(z)$ itself will appear in expressions for the aberration coefficients. For compactness we shall omit the explicit references to z in R and T^* in what follows.

We shall integrate the equation using equal steps, h, in z and so write for the values of R at adjacent steps

$$R_{n\pm 1} = R_n \pm hR'_n + \frac{h^2}{2}R''_n \pm \frac{h^3}{6}R'''_n + \frac{h^4}{24}R_n^{(4)} \pm \cdots$$

and rearrange this to get

$$\frac{R_{n+1} - 2R_n + R_{n-1}}{h^2} = R''_n + \frac{h^2}{12}R_n^{(4)} + \cdots.$$

$R''_n = -(T^*R)_n$ and the fourth differential of R_n can be written in terms of the second differential of $(T^*R)_n$ as

$$R_n^{(4)} = \frac{\mathrm{d}^2(-T^*R)_n}{\mathrm{d}z^2}$$
$$= -\frac{(T^*R)_{n+1} - 2(T^*R)_n + (T^*R)_{n-1}}{h^2}$$

using an exactly similar expansion of T^*R, but neglecting the fourth-order term in the expansion, giving

$$\frac{R_{n+1} - 2R_n + R_{n-1}}{h^2} = -(T^*R)_n - \frac{h^2}{12}\frac{[(T^*R)_{n+1} - 2(T^*R)_n + (T^*R)_{n-1}]}{h^2}$$

and, rearranging this equation, we finally have

$$R_{n+1} = \frac{2R_n - R_{n-1} - \frac{h^2}{12}[(T^*R)_{n-1} + 10(T^*R)_n]}{[1 + \frac{h^2}{12}T^*_{n+1}]}. \tag{4.3}$$

This equation is known as Numerov's algorithm† and it is very effective within its limitations. Two values of $R(z)$ are required to start the process, but this is not a problem if the starting point is in a field-free region. If this is not the case, some other method has to be used to start the process,

---

† It is also known as the Manning–Millman and the Fox–Goodwin methods.

and a Runge–Kutta method would normally be used. The step size has to be chosen carefully as there is no easy way to adjust it dynamically, since the particle velocity is not calculated at any stage. This means that a step size suitable for the worst part of the trajectory has to be used throughout.

4.3.2.1 *The Runge–Kutta method applied to the Picht equation.* As usual, the Runge–Kutta method requires parameters to be calculated in the middle of the integration steps. If the values of T^* are known at intervals $(h/2)$, the Runge–Kutta method must use a step size of h while the Numerov method can use the smaller step giving it a 16-fold advantage in precision. The expressions for the reduced radius R and its gradient R' are

$$R_{n+1} = R_n - R_n \frac{h^2}{6}(T_n^* + 2T_{n+1/2}^* - \frac{h^2}{4}T_n^* T_{n+1/2}^*) + R'_n(1 - \frac{h^2}{6}T_{n+1/2}^*)$$

$$R'_{n+1} = -R_n \frac{h}{6}(T_n^* + 4T_{n+1/2}^* - \frac{h^2}{2}T_n^* T_{n+1/2}^*) + R'_n(1 - \frac{h^2}{3}T_{n+1/2}^*)$$
$$- \frac{h}{6}T_{n+1}^* R_{n+1}.$$

Notice that it is necessary to calculate R_{n+1} before R'_{n+1}.

4.4 Real and asymptotic cardinal points

Particle lenses differ from those for light in having no abrupt boundaries between regions of different refractive index, so there is no clear-cut point at which the lens begins or ends. Though the ray paths are asymptotically straight away from the lens, a particle emerging from the lens field may cross the axis before this asymptotic region is reached. If the particle had approached the lens parallel to the axis, this crossing point will be a focal point and the mid-focal distance will be well defined, but how should the focal length be defined? In Chapter 1 the focal length was defined as the distance of the focal point from a principal plane which was itself defined by the intersection of asymptotes. This is not appropriate to the present case and we must define a *real* principal plane in terms of the incident asymptote and a tangent to the ray path at the focal point. These real cardinal points are of limited use in the analysis of lens behaviour because for an object not at infinity the image will be formed beyond the focal point and therefore closer to the asymptotic region, if not fully in it. Fortunately, the distinction is necessary only for lenses which are stronger than are commonly used except, perhaps, for microscope objectives.

4.4.1 The determination of the cardinal points by ray tracing

If the path of a ray starting some 3 to 4 diameters from the lens is traced, using any of the methods outlined in section 4.3, the position of the focal point can be found by noting the values of z between which the radius, r, or the reduced radius, R, changes sign and then interpolating to find the position of the crossing point. This will be the true focal point, but unless the lens is very strong indeed, it will be very close to the asymptotic focal point. If the ray path has not crossed the axis by the end of the integration, the focal point can be located by extrapolating the last few points calculated. The corresponding principal point is found from the slope of either the tangent to the path at the focal point or of the asymptote. With a good potential the effects of spherical aberration can be seen by changing the off-axis distance of the incident ray, and the spherical aberration coefficient can be determined. If the trace is done by integrating the Picht equation, it is not necessary to convert back to the real radius as both r and R are zero at the crossing point and the asymptote is only reached in a region of constant potential. Nonetheless, it is interesting and instructive to make the conversion as the approach to the asymptotes has a rather different appearance.

4.5 Windows and pupils

While it is the paths of individual rays which are calculated, measurements are made with *beams* of particles and these have finite lateral and angular extent. These beams must be defined by suitable apertures, and in figure 4.5 we illustrate some of the terms and parameters involved. The beam is defined by two apertures, called conventionally the window and the pupil. The window is regarded as the object on which the lens acts to produce an image and the distinction is sometimes made between the object window and the image window. The pupil, or angle stop, defines the range of angles which may enter the lens from any point in the window. Naturally the lens produces an image of the pupil also, and the position and size of the image pupil are important parameters of an optical system. In figure 4.5 the lens is represented as a thin lens with coincident principal planes purely to avoid complexity in the diagram. The lateral dimensions are grossly exaggerated. The upper part of the diagram shows the paths of a number of rays from the bottom of the object window and a point within it which pass through the object pupil at its upper and lower edges and also through the centre of the pupil. In the image space the rays from each point in the object window cross at the corresponding point in the image window and those rays which passed through particular points of the object pupil pass through the corresponding parts of the image pupil. Two sets of angles are indicated: α and θ. α is known as the beam angle

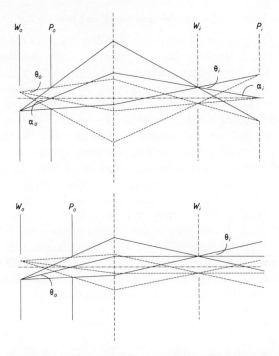

Figure 4.5   Diagram to illustrate the formation of the image of an entrance, or object, window, W_o, by a bundle of rays limited by an entrance, or object, pupil, P_o. The semi-angular divergence of the bundle of rays is specified by the *pencil angle, θ_o* and the maximum inclination of the central ray of the bundle to the axis in the plane of the entrance pupil is the *beam angle, α_o*. The lens forms images of the object-side window and pupil, and the lower diagram shows the special case where the object-side pupil is at the first focal point, giving an image pupil at infinity and zero beam angle on the image side.

and is the symbol used for the angle between the principal rays in the construction of figure 1.3. It is the angle subtended at the centre of the pupil by the radius of the window and is therefore the angle at which the central ray from the edge of the window crosses the axis in the plane of the pupil. θ is the pencil angle, describing the angular spread of a pencil of rays from a point in the window and is the angle subtended at the centre of the window by the radius of the pupil. Subscripts distinguish the object and image space values of these angles. If the radii of the window and pupil are written as r_w and r_p, respectively, and the distance between them is l then the angles can be expressed as

$$\alpha = \frac{r_w}{l} \qquad \theta = \frac{r_p}{l}.$$

The lower part of figure 4.5 shows a special case. The object pupil is at the first focal point of the lens and is therefore imaged at infinity. The image-side beam angle is now zero and the total angular divergence of the beam, which can be written as $(|\alpha| + |\theta|)$ is a minimum.

4.5.1 The energy-add lens

If a real image of a window is formed at the first principal plane of a lens, the emergent rays appear to come from a virtual image of the same size at the second principal plane. These rays may be focused to give a final, real image. The simplest practical system consists of four cylinders. The central two cylinders form the energy-add lens proper, and exploit the fact that the positions of the principal planes of a two-cylinder lens do not change much over a fairly wide range of voltage ratios. The lens formed by the first two cylinders produces the image at the first principal plane of this central lens, and that formed by the last two cylinders gives the final image. The voltage ratios V_2/V_1 and V_4/V_3 are held constant, but the ratio V_3/V_2 is changed, *adding* energy $e(V_3 - V_2)$ to the particle. This apparently simple method of maintaining the position of the final image over a range of overall voltage ratio, V_4/V_1, has major drawbacks for all but the crudest applications. While the imaging of the entrance window may be quite well controlled, there is no control over the position and size of the pupil image formed by the first lens. The action of the second and third lenses will only make matters worse, most probably leading to excessive beam angles at the final image. The presence of an intermediate image is also likely to cause increased aberrations. The simple multi-element lenses described in the next section are much better behaved.

4.6 Multi-element lenses

4.6.1 Zoom lenses

A lens having two elements allows a conjugate focus condition to be met in just two ways, with an accelerating or a decelerating voltage ratio. The addition of a third element allows the condition to be met over a wide range of overall voltage ratios by a suitable choice of the potential of the third, centre, electrode. The focal locus for one such lens is shown in figure 4.6. The magnification of the lens can only be controlled in the most general fashion by, for example, selecting the lengths of the cylinders to give particular values at a limited number (three, at the most) of points on the locus. In order to make a true zoom lens, one in which the magnification for a given pair of conjugates can be controlled independently of the overall voltage ratio, a fourth element is necessary. It is difficult to obtain much flexibility with aperture lenses because the electrodes have to be comparatively close together but, with cylindrical electrodes, the distance between

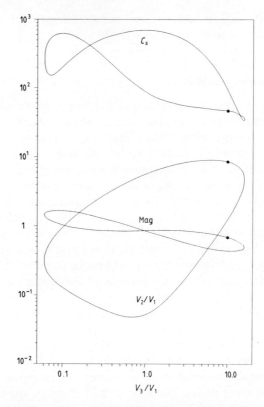

Figure 4.6 Focal locus for a three-cylinder lens having elements of length 3.0, 0.75 and 2.5 diameters. The variation of the magnification and of the coefficient of spherical aberration round the locus are also shown. A ● identifies corresponding points on the three loci.

the two outer electrodes, which determines the overall voltage ratio, can be usefully large. The two central electrodes must have potentials which combine in some way to satisfy the focus condition, but this can be done in many ways and one might consider the two potentials to have some average magnitude (though not a simple arithmetic mean) and also a 'centre of action' which can be moved to the left or right by changing their *ratio*. Figure 4.7 illustrates this for two four-cylinder lenses with symmetric conjugates separated by 6 diameters and with their two central elements of length D in one case and $D/2$ in the other. This figure shows the magnifications which may be obtained for an overall voltage ratio of unity. The lens with the longer central elements offers a range of magnifications from 0.67 to 1.5 while the other will only cover from 0.87 to 1.15. Notice that the extreme magnifications are obtained with both centre electrodes at higher potentials than the outer ones and not, as one might have expected, with one of the centre elements at the outer potential. It is left to the reader to investigate (using the **LENSYS** program) other possible pairs of values giving the same magnifications and to deduce why they are not good combinations.

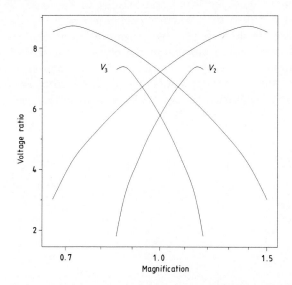

Figure 4.7   The potentials of the two centre elements of four-cylinder lenses of total length 6 diameters required to give a range of magnifications. The upper curves relate to a lens with centre elements each of length 1 diameter, and the lower curves to one with lengths of 0.5 diameter. The potentials of the first and fourth elements are each equal to 1.

The greater range of magnification obtained with the longer central elements raises the question of further increase. However, just extending the lengths of these two elements will not help a great deal. A much better approach is to place a fifth electrode in the middle of the system, and for the moment one can think of this merely as a 'spacer' keeping the two magnification-controlling electrodes a good distance apart! Before discussing this way of obtaining zoom behaviour on a par with that obtainable with quite modest cameras, it will be helpful to examine a somewhat specialised five-element cylinder lens.

4.6.2   The afocal lens

We consider in this section a *combination* of two lenses which has an interesting and useful property. Figure 4.8 shows the optics of this combination schematically and illustrates the construction, using principal rays, of the image of an object to the left of the first lens of the pair. The essential point of the lens combination is that the second focal point of the first (left hand) lens coincides with the first focal point of the second (right hand) lens. This means that the first principal ray, drawn from the top of the object parallel to the axis, passes through this common focal point and emerges from the second lens again parallel to the axis. The immediate consequence of this is that, wherever the object may be, the image will be the same size: in other words, the magnification of the lens combination does not depend on the position of the object. The second principal ray is drawn through the first focal point of the left hand lens, emerges from that lens parallel to the axis and is then deviated to emerge from the right hand lens as though from the second focal point of that lens.

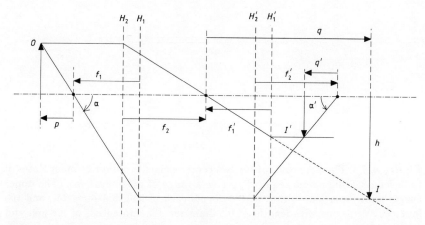

Figure 4.8 The construction of the image of a given object by two lenses having a common focal point in the region between them.

The focal lengths and the object and image distances are indicated on the figure and we use Newton's relation to examine their relationship. The first lens would produce an intermediate image I of O a distance q from the common focal point such that $q = f_1 f_2/p$ where p is the distance of the object from the first focal point of the left hand lens. This image lies beyond the first focal point of the second lens and is imaged by that lens as I', a distance q' from the second focal point of this lens such that $q' = f_1' f_2'/q$. In terms of the position of the original object $q' = (f_1' f_2'/f_1 f_2)p$. If the two lenses are identical then we have the interesting result that $q' = p$ and the final image is the same distance from the second focal point of the right hand lens as the object is from the first focal point of the left hand lens. The fact that rays entering parallel to the axis emerge parallel to the axis means that the lens system has no focal points — it is *afocal* or telescopic.

Remembering that F_1 is a negative quantity, the distance between the object and the final image is $2(F_2 - F_1)$, and this must be held constant, independent of the overall voltage ratio, if the special property of this lens is to be maintained. The values of F_1 and of F_2 will change, of course, and if the difference is not to change, each lens must consist of three electrodes. The last element of the left hand lens and the first element of the right hand lens are at the same potential (otherwise there would be further lens action in the middle of the system) so the entire lens consists of five elements. Because the two lenses are identical

$$\frac{V_2}{V_1} = \frac{V_4}{V_3} \qquad \text{and} \qquad \frac{V_3}{V_1} = \frac{V_5}{V_3}$$

and therefore

$$\frac{V_3}{V_1} = \left(\frac{V_5}{V_1}\right)^{1/2}.$$

The magnification of the lens can be calculated from the figure and is $M = f_1'/f_2 = f_1/f_2$ for identical lenses. The angular magnification is given by

$$M_\alpha = \frac{\alpha'}{\alpha} = \left(\frac{h}{f_2'}\right) \bigg/ \left(\frac{h}{f_1}\right) = \frac{f_1}{f_2'} = \frac{f_1}{f_2} = M.$$

The magnification, angular magnification and the overall voltage ratio are related by the law of Helmholtz and Lagrange

$$M M_\alpha \left(\frac{V_5}{V_1}\right)^{1/2} = 1$$

so

$$M = M_\alpha = \left(\frac{V_5}{V_1}\right)^{-1/4}.$$

Apart from the trivial case of a lens which is electrically and mechanically symmetric, this is the only lens for which the magnification can be calculated without any knowledge of the potential distribution; it depends only on the *identity* of the two component lenses.

An important application of this lens is in the production of a well defined particle beam. The images of an object-side window and pupil will have the same axial separation and be magnified by the same factor, so if the object-side apertures are of equal diameter, the particle beam will be bounded by a cylindrical surface in a region of the image space though no apertures are present there, reducing the chance of unwanted collisions of the beam with the metalwork. The beam angle and the pencil angle will be equal in this situation.

4.6.3   The zoom afocal lens

If we remove the constraint on the two component lenses of figure 4.8, but maintain the same distance between conjugates, the lens system will not, in general, remain afocal. It does then, however, have the desirable property of the zoom lens. The magnification is controlled by the relative strengths of the first and second lenses in just the same way as is done by the two centre electrodes of a four-cylinder lens. For the moment we keep the potential of the centre electrode at the geometric mean of those of the first and fifth electrodes and show in figure 4.9 (by broken curves) the values of the potentials of the second and fourth electrodes required to give a range of magnifications. The particular lens geometry consists of two lenses, each of three cylinders having lengths of 1.25, 0.5 and 1.25 diameters joined end to end to give a five-element lens with a centre element 2.5 diameters long and a total length of 6 diameters, the same as the two four-cylinder lenses shown in figure 4.7. The very substantial increase in the range of magnifications is apparent.

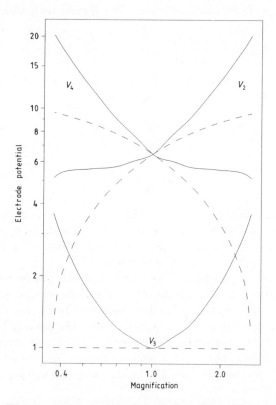

Figure 4.9   The electrode potentials required to give a range of magnification for a five-cylinder lens of total length 6 diameters with $V_5 = V_1 = 1$. Two sets of data are shown; the broken curves are for the case of $V_3 = 1$, and the solid curves are for V_3 adjusted to give a pupil magnification equal to the window magnification.

Only four elements are necessary for a lens to have a zoom capability over a range of voltage ratios, and the present lens has, therefore, an extra element which might be used to control another parameter of the lens. There is more than one possibility, but in the present context the control of a second conjugate pair would allow the lens to be again afocal. The effect of adjusting the potential of the centre electrode (with compensating adjustments of those of the second and fourth) is illustrated in figure 4.10 for this same 6 diameter length lens. The potentials have been adjusted to give a window magnification of -1 and the figure shows the magnification of a pupil located a distance $D/2$ to the right of the window as a function of the voltage ratio V_3/V_1 for a number of values of the overall voltage ratio, V_5/V_1. The points marked ● correspond to $V_3/V_1 = \sqrt{V_5/V_1}$. Notice that, for both accelerating and retarding lenses, V_3 has to be increased to make the pupil magnification equal that of the window and so make the lens afocal. For an overall voltage ratio of 1, the potentials of the three centre electrodes are shown also in figure 4.9 (by full curves) for this zoom afocal configuration.

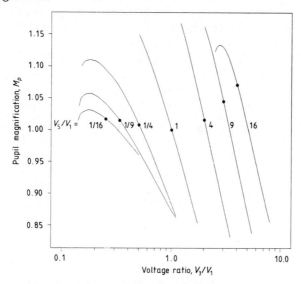

Figure 4.10   The pupil magnification of a five-cylinder lens as a function of the potential of the centre element for various values of the overall voltage ratio, V_5/V_1. The lens is operated to give a window magnification of -1 and the object pupil is a distance $D/2$ from the object window.

Chapter 5

Aberrations

5.1 Spherical aberration coefficients

Particle lenses are subject to the same optical aberrations as photon lenses and the effects can be much worse because it is not possible to use materials of different dispersion to reduce chromatic aberration nor to grind surfaces of special forms to reduce spherical and other aberrations. The handling of particle beams is usually an axis-centred problem so spherical aberration is normally much more serious than the off-axis aberrations, such as coma, astigmatism and distortion. We shall therefore consider only rays which cross the axis at some point. Figure 5.1 is a schematic diagram showing the emergent asymptote corresponding to a ray incident from an axial object point. For meridional rays, the relationship of the radial positions and slopes of the rays at the first and second focal planes of the lens can be expressed as [11]

$$r_2' = -\frac{r_1}{f_2} + m_{13}r_1'^3 + m_{14}r_1'^2\frac{r_1}{f_2} + m_{15}r_1'\left(\frac{r_1}{f_2}\right)^2$$
$$+ m_{16}\left(\frac{r_1}{f_2}\right)^3 + \cdots \tag{5.1a}$$

$$-\frac{r_2}{f_1} = r_1' + m_{23}r_1'^3 + m_{24}r_1'^2\frac{r_1}{f_2} + m_{25}r_1'\left(\frac{r_1}{f_2}\right)^2$$
$$+ m_{26}\left(\frac{r_1}{f_2}\right)^3 + \cdots \tag{5.1b}$$

remembering that f_1 is negative.† The coefficients, m_{ij}, are dimensionless and also negative. Higher powers of r and r', beginning with six terms of

---

† A number of the equations in this chapter differ from their counterparts in other texts as a consequence of our cartesian sign convention.

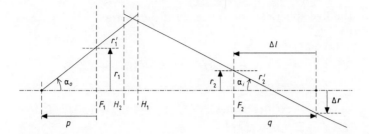

Figure 5.1 Definition of the ray parameters for the description of the meridional aberration coefficients.

fifth order for which the coefficients are usually written as q_{ij}, have been ignored for the time being.

The coefficient of spherical aberration for a lens operated with magnification M, $C_s(M)$, is defined by

$$\Delta r = -MC_s(M)\alpha_o^3$$

where $\alpha_o = -r_1'$. From figure 5.1 we can write $r_1 = -pr_1'$, and $\Delta r = r_2 + qr_2'$ for the distance off-axis at which the ray crosses the Gaussian image plane. The object and image distances can be written in terms of the focal lengths and the magnification as $p = -f_1/M$ and $q = -Mf_2$. Substitution of these expressions into equation (5.1) gives the following relationship between the coefficient of spherical aberration and the meridional coefficients:

$$C_s(M) = -m_{13}f_2 - (m_{14} + m_{23})(f_1/M) - (m_{15} + m_{24})(f_1/M)^2/f_2$$
$$- (m_{16} + m_{25})(f_1/M)^3/f_2^2 - m_{26}(f_1/M)^4/f_2^3. \quad (5.2)$$

Each term of equation (5.2) has a positive value so no cancellation is possible. C_s has the dimensions of Δr, but it is a common convention to express both in terms of the diameter of the lens, and all the values presented in this book and in the **LENSYS** program are strictly of C_s/D. Equation (5.2) shows that $C_s(M)$ could be expressed as a fourth-order polynomial in the object distance, $p = -f_1/M$, but it is more usual to write it in terms of the magnification, M,

$$C_s(M) = C_{s_0} + C_{s_1}/M + C_{s_2}/M^2 + C_{s_3}/M^3 + C_{s_4}/M^4 \quad (5.3)$$

where the C_{s_i} are properties of the lens itself.

There are certain well established relationships between some of the eight meridional aberration coefficients

$$m_{14} = 3m_{23} + 1.5 \qquad m_{15} = m_{24} \qquad m_{25} = 3m_{16} + 1.5 \qquad (5.4)$$

and so the five spherical aberration coefficients are

$$C_{s_0} = -m_{13}f_2$$
$$C_{s_1} = -(m_{14} + m_{23})f_1 \quad = -(4m_{23} + 1.5)f_1$$
$$C_{s_2} = -(m_{15} + m_{24})f_1^2/f_2 \quad = -2m_{24}f_1^2/f_2 \tag{5.5}$$
$$C_{s_3} = -(m_{16} + m_{25})f_1^3/f_2^2 \quad = -(4m_{16} + 1.5)f_1^3/f_2^2$$
$$C_{s_4} = -m_{26}f_1^4/f_2^3.$$

If the object is at infinity, $r_1' = 0$ and the magnification is therefore zero. It is then necessary to use equation (5.1b) directly. Including the relevant fifth-order term, $q_{26}(r_1/f_2)^5$, we have

$$\Delta r = r_2 = -m_{26}f_1(r_1/f_2)^3 - q_{26}f_1(r_1/f_2)^5.$$

The third-order term can be written as

$$\Delta r = C_{s_4}(r_1/f_1)^3. \tag{5.6}$$

5.1.1   The axial displacement

The aberrated ray crosses the axis a distance Δl before the Gaussian image point where

$$\Delta l = -\frac{\Delta r}{r_2'} = -MC_s\frac{\alpha_o^3}{\alpha_i} = -\frac{M}{M\alpha}C_s\alpha_o^2 = M^2C_s\alpha_o^2\frac{f_2}{f_1} = M^4C_s\alpha_i^2\left(\frac{f_2}{f_1}\right)^3 \tag{5.7}$$

and where $\alpha_i = -r_2'$ is the beam angle in the image space.
For parallel input $M = 0$ so $M^4C_s = C_{s_4}$ and, using $\alpha_i = r_1/f_2$,

$$\Delta l = \left(\frac{f_2}{f_1}\right)^3 C_{s_4}\alpha_i^2 = \frac{f_2}{f_1}C_{s_4}\left(\frac{r_1}{f_1}\right)^2.$$

If we wish to include the fifth-order term without excessive algebra, we can write

$$r_2' = -r_2/\Delta l = -r_1/(f_2 + \Delta l)$$

giving

$$r_2 = \frac{r_1\Delta l}{f_2 + \Delta l}$$

from which we find the following working expression

$$\left(\frac{f_2}{f_1}\right)\frac{\Delta l}{f_2 + \Delta l} = -m_{26}\left(\frac{r_1}{f_2}\right)^2 - q_{26}\left(\frac{r_1}{f_2}\right)^4$$

but this is not likely to give a very good value for the fifth-order coefficient, though it will improve that of the third order.

5.1.2   The disc of minimum confusion

If the object is a point source of particles, the 'image' in the Gaussian plane will be a round spot with a radius determined by the pencil angle and the aberration coefficients of the lens. Rays which have passed close to the limits of the entrance pupil will contribute to the outer parts of this spot, having crossed the axis somewhat closer to the lens. Between the Gaussian plane and the point on the axis where these marginal rays are focused, there will be some rays which cross the axis going upwards and some going downwards, with the former being more steeply inclined, and the spot formed by the envelope of all the rays will have a minimum radius somewhere between these points. This spot is known as 'the disc of minimum confusion'. Figure 5.2 shows the marginal ray and a general ray which intersect a distance h from the axis; the axial position of the intersection is a distance l from the Gaussian focus, G. We can express h in two ways

$$h = \alpha_g(z_g - l) = \alpha_m(z_m - l)$$
$$A\alpha_g^3 - l\alpha_g = A\alpha_m^3 - l\alpha_m$$

so

$$l = A(\alpha_m^2 + \alpha_m\alpha_g + \alpha_g^2)$$

and

$$h = -A(\alpha_m^2\alpha_g + \alpha_m\alpha_g^2)$$

where $A = C_{s_4}(f_2/f_1)^3$ from equation (5.7), and is negative. We find the minimum value of h by differentiation

$$dh/d\alpha_g = -A(\alpha_m^2 + 2\alpha_m\alpha_g) = 0 \qquad \text{if} \quad \alpha_g = -\alpha_m/2$$

and this leads to the radius of the disc of minimum confusion being

$$h_{min} = A\frac{\alpha_m^3}{4}$$

and its position as

$$l = \frac{3}{4}A\alpha_m^2.$$

The radius is a quarter of that of the Gaussian plane spot and it is three quarters of the way from the Gaussian image to the most aberrated image point.

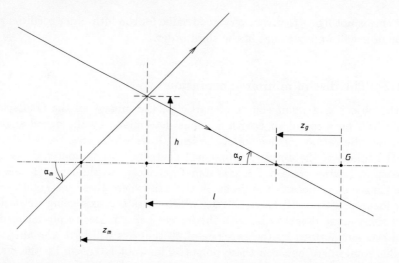

Figure 5.2   Aberrated rays in the region of the disc of minimum confusion.

5.2   Off-axis aberrations

Figure 5.3 is similar to figure 5.1, but indicates the asymptotes to a ray from an object point, a distance r_0 from the axis, which crosses the first focal plane at a radial distance r_1. The path of the ray is still in a plane containing the axis; we shall not be considering skew rays. Using equations (5.1) again, we write the position of the ray at the image plane as the sum of the geometric, Gaussian distance, Mr_0, and an aberration term which can be written as the sum of terms involving $r_0^3, r_0^2 r_1, r_0 r_1^2$ and r_1^3

$$
\begin{aligned}
\Delta r = Mr_0 &+ r_0^3\left[M^4 m_{13}\frac{f_2}{f_1^3} + M^3 m_{23}\frac{1}{f_1^2}\right] \\
&- r_0^2 r_1\left[3M^4 m_{13}\frac{f_2}{f_1^3} + M^3(m_{14}+3m_{23})\frac{1}{f_1^2} + M^2 m_{24}\frac{1}{f_1 f_2}\right] \\
&+ r_0 r_1^2\left[3M^4 m_{13}\frac{f_2}{f_1^3} + M^3(2m_{14}+3m_{23})\frac{1}{f_1^2}\right. \\
&\qquad\qquad \left. + M^2(m_{15}+2m_{24})\frac{1}{f_1 f_2} + M m_{25}\frac{1}{f_2^2}\right] \\
&- r_1^3\left[M^4 m_{13}\frac{f_2}{f_1^3} + M^3(m_{14}+m_{23})\frac{1}{f_1^2} + M^2(m_{15}+m_{24})\frac{1}{f_1 f_2}\right. \\
&\qquad\qquad \left. + M(m_{16}+m_{25})\frac{1}{f_2^2} + m_{26}\frac{f_1}{f_2^3}\right].
\end{aligned}
$$

The terms in r_0^3 describe the distortion of the image because they imply a magnification which depends on the position of the point in the object but not on the lens aperture. From equation (5.5) we see that C_{s_0} and C_{s_1} describe this aberration. The terms in $r_0^2 r_1$ describe the combined effects of astigmatism and curvature of field and those in $r_0 r_1^2$ describe coma. The terms in r_1^3 describe spherical aberration and transform to the form of equation (5.2) on writing $r_1 = -(f_1/M)\alpha_o$. If $r_0 = r_1$, which implies $M = 0$, only the term in m_{26}, equivalent to C_{s_4} remains and equation (5.6) is recovered. Notice that m_{26} appears only in the coefficient of r_1^3.

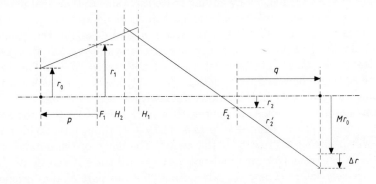

Figure 5.3   Definition of the ray parameters for the description of the off-axis aberrations of a lens.

5.3   The calculation of aberration coefficients

If the potential distribution throughout the lens is known with sufficient accuracy, rays will follow their correct paths and, if they are not paraxial, will show the effects of the lens aberrations. In principle, then, all that is needed is careful ray tracing to find the intercept of the ray at the Gaussian image plane. This raises the first problem: where is the Gaussian image plane? It is necessary to trace more than one ray and extrapolate to zero α_o or r_1 as appropriate and it is tempting to use the axial displacement directly. If only the axial potential is known, then calculations must be based on the effect of the terms omitted from the axial equation of motion.

5.3.1   Direct ray tracing

This section provides an opportunity to describe the operation of the **SIMION** package introduced briefly in section 3.3.4. We shall consider the behaviour of rays initially parallel to the axis in a two-cylinder lens with a voltage ratio of 10 : 1, looking at both the accelerating and decelerating modes. First we must define the electrode geometry and **SIMION** allows the user to specify one of three basic geometries:

Non-symmetrical planar

Symmetrical planar

Cylindrical symmetrical.

In the latter two cases **SIMION** asks if the mirror image should be displayed and the response will normally be 'yes'. You are then asked for the maximum electrode potential used. This is because the lattice points associated with electrodes are flagged by adding twice this value to their real potential and a high enough value must be given here to avoid ambiguity. The final questions ask for the size of the array, $x \times y$, and details of the geometry itself. In the present example, because of the symmetry, the dimensions are length × radius. The product xy must not exceed 16 000. The appropriate response here will be something like $321, 49$ which will allow a total cylinder length of 4× the diameter (80) with a wall thickness of 9. This is just about long enough to cope with a 10 : 1 lens. Note that the actual coordinates will run from $0, 0$ to $(x - 1), (y - 1)$ though the axial potential will be listed as 'row 1' of the array! The entry of the details of the geometry is straightforward and not very critical since it can be modified interactively later. The electrode potentials form part of the 'geometry' data and, for reasons which will be clear in a moment, should be set to values of '1' and '2' (and '3','4',..., for systems with more electrodes). The file should then be saved as (e.g.) '**LENS.PA#**'. The extension is important as this is the 'Master Potential Array' and the next stage is to run the program **RPA.EXE** ('Refine Potential Array'), which is part of the **SIMION** package. This makes a series of five-point relaxation analyses of the array using user-specified values for the names of the electrodes, the acceleration parameter (which **SIMION** identifies as $(g - 1)$ in the notation of Chapter 3), and the target precision. **RPA** stores its output as files **LENS.PA0**, **LENS.PA1** and **LENS.PA2** which are each the size of **LENS.PA#**, but contain specific data. **LENS.PA1**, for example, is the potential array found by setting the potential of the first electrode to 10 000 volts and those of all the other electrodes to zero. Re-entering **SIMION**, the next stage is to load **LENS.PA0** as an 'OLD' file and run the Fast Adjust option: this asks for the *actual* potentials of the two electrodes and generates the potential distribution corresponding to these values by scaling and superposing those in the **LENS.PAn** files. This scaled array should now be saved as **LENS.PA** and it may be recalled as an 'OLD' file later for actual use. This whole process takes about an hour, but the time needed to recalculate for a different set of potentials is only a few seconds.

The **RPA.EXE** program does not produce a man-readable list of potentials, but there is a way of obtaining such a list. In **SIMION** itself there is a Refine option which makes a five-point analysis just like **RPA.EXE**, but on one single array rather than on the set of arrays implied in a *.PA# file. It is usually much more effective to use **RPA.EXE** with subsequent scaling by the Fast option, but if you run the Refine option on an already solved potential, it will find that the solution has already converged and will write the whole potential array in its log file, from which you may extract any parts required using a simple text processor.

SIMION will calculate and plot equipotential lines on the electrode system and figures 1.2(*b*) and 4.3 were produced in this way. The principal objective in using SIMION is to examine the ray paths and, to enable this, there is a Trajectory option within which the user may specify the actual size of the system (in millimetres), scale the potentials and specify the charge-to-mass ratio of the particle. Naturally, the starting positions and slopes can be chosen and grouped as initial values plus increments. The output of the Trajectory option is both graphical and a list of data written to a log file. Figure 5.4 shows rays traced in both the accelerating and decelerating lenses. The much greater effect of spherical aberration in the retarding lens is apparent. For the purposes of the present analysis, the important parameters of the trajectory are those indicated in figure 5.5.

(*a*)

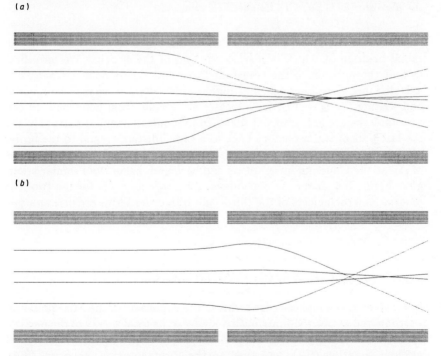

(*b*)

Figure 5.4 Rays traced in a two-cylinder lens having voltage ratios of (*a*) 10 : 1, and (*b*) 1 : 10. The aberrations of the retarding lens are so large that a ray similar to the outermost ray of figure (*a*) would strike the inner surface of the lens well before the end of the diagram. These diagrams were produced by the SIMION program.

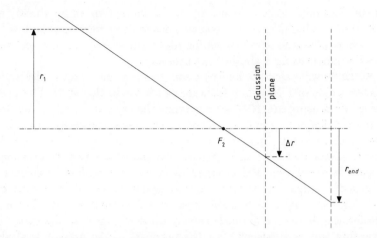

Figure 5.5   The **SIMION** program provides values of r_{end} and the position of the *aberrated* focal point F_2 for specified values of r_1.

SIMION records the position at which the ray crosses the axis and the radial position at which it reaches the end of the system. We have first to find the position of the Gaussian focus and this is done by using off-axis distances small enough that higher-order aberrations may be ignored and writing $l = l_0 - Ar_1^2$. This locates the second focal point and we can now find the second principal plane from similar triangles, extrapolating the last part of the trajectory back to a radial distance equal to the initial off-axis distance. These two data give the second focal length, and values of Δr at the Gaussian image plane can be found, again from similar triangles. Figure 5.6 shows the dependence of $\Delta r / f_1$ on r_1 / f_2 for the two lens modes. The coefficients of the third- and fifth-order terms are m_{26} and q_{26}, respectively.

5.3.2   Perturbation of the paraxial solution

In Chapter 2, the higher-order terms in the expansion of the axial potential were ignored as was the contribution made to the total energy of the particle by its radial velocity. These simplifications led to the paraxial equation of motion, equation (2.3), and then to the Picht equation, equation (2.4). In order to describe the non-paraxial behaviour, it is necessary to replace these terms. The equation of radial motion becomes

$$m\frac{\mathrm{d}^2 r}{\mathrm{d}t^2} = eE_r = e\left[\frac{r}{2}V'' - \frac{r^3}{16}V^{(4)}\right]$$

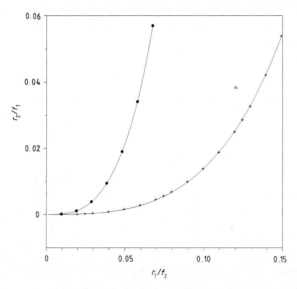

Figure 5.6 Graphs of $\Delta r/f_1$ against r_1/f_2 for the lenses shown in figure 5.4. The curves represent $(\Delta r/f_1) = 12.6(r_1/f_2)^3 + 176(r_1/f_2)^5$ (+), and $(\Delta r/f_1) = 153(r_1/f_2)^3 + 3610(r_1/f_2)^5$ (•) for the accelerating and retarding lenses, respectively.

and the total energy of the particle has to be written as

$$\frac{1}{2}m\left[\left(\frac{dz}{dt}\right)^2 + \left(\frac{dr}{dt}\right)^2\right] + e\left[V - \frac{r^2}{2}V''\right] = 0.$$

The equation of motion is then

$$\frac{d^2r}{dz^2} + \frac{1}{2}\frac{V'}{V}\frac{dr}{dz} + \frac{r}{4}\frac{V''}{V} = \left[\frac{1}{32}\frac{V^{(4)}}{V} - \frac{1}{16}\left(\frac{V''}{V}\right)^2\right]r^3 \qquad (5.8)$$
$$+ \frac{1}{8}\left(\frac{V'''}{V} - \frac{V''V'}{V^2}\right)r^2r' - \frac{1}{4}\frac{V''}{V}rr'^2 - \frac{1}{2}\frac{V'}{V}r'^3.$$

To solve this equation, we write $r = r_p + r_a$ where r_p is the solution to the equation of paraxial motion and r_a is a *small* perturbation representing the aberration due to the use of better approximations to the real potential. r_a is small enough that only first-order terms need to be retained. Each term of the left hand side of equation (5.8) splits into one containing r_p and one containing r_a. Those in r_p sum to zero, because they are just the terms of the paraxial equation. The terms on the right hand side involve the third power of r and so only the r_p part survives, giving an equation for the perturbation, r_a,

$$\frac{d^2 r_a}{dz^2} + \frac{1}{2}\frac{V'}{V}\frac{dr_a}{dz} + \frac{r_a}{4}\frac{V''}{V} = \left[\frac{1}{32}\frac{V^{(4)}}{V} - \frac{1}{16}\left(\frac{V''}{V}\right)^2\right] r_p^3 \tag{5.9}$$

$$+ \frac{1}{8}\left(\frac{V'''}{V} - \frac{V''V'}{V^2}\right) r_p^2 r_p' - \frac{1}{4}\frac{V''}{V} r_p r_p'^2 - \frac{1}{2}\frac{V'}{V} r_p'^3.$$

Solutions of the homogeneous equation, obtained by setting the right hand side of equation (5.9) equal to zero, are known: they are just the general solutions of the Gaussian ray equation discussed in Chapter 1. We can therefore write the perturbation as $\zeta(z)r_1(z) + \xi(z)r_2(z)$ where we stress the variability of ζ and ξ, reflecting the non-zero right hand side. The problem reduces to the calculation of $\xi(z_i)$ where z_i is the position of the Gaussian image. After a lot of algebra, two integrations by parts and transformation of the radial variable to the reduced radius, $R = rV^{1/4}$, the coefficient of spherical aberration can be written as

$$C_s(M) = \frac{1}{64V_0^{1/2}} \int_{-\infty}^{\infty} \frac{R^4}{V^{1/2}} \left[(3T^4 + 5T'^2 - \tfrac{11}{2}T^2T') + 4TT'\frac{R'}{R}\right] dz \tag{5.10}$$

where V_0 is the potential at the lower limit of integration.† In this equation $R(z)$ corresponds to the ray which passes through the axial object point with unit slope. Every variable, bar one, in this equation will have been determined in the course of calculating the paraxial image position. The missing variable is T' and it is a trivial matter to find this. We noted earlier that the sign of T is not important to the calculation of the paraxial focus as it appears as T^2 in the Picht equation, but the presence of T' here shows that it does matter for the calculation of the aberrations. Though the limits of integration are marked as $\pm\infty$, in practice a much shorter interval will suffice.

Figure 5.7 shows the integrand of equation (5.10) for the cases of a three-aperture lens, used asymmetrically, and a five-cylinder lens used in the afocal mode. The range of axial position over which the integrand is significantly greater than zero is appreciably less than the distance between the conjugate points. For a two-element lens, we might expect the integrand to show two maxima, one where R is large and the other where R' is large. In the three-element lens, especially with aperture electrodes, two of the four possible maxima run into each other, but the cylinder lens shows two

---

† The use of the reduced radius has transformed a term in $(V/V_0)^{1/2}$ into $(1/V_0V)^{1/2}$. Almost always, the potentials will be expressed in terms of V_0.

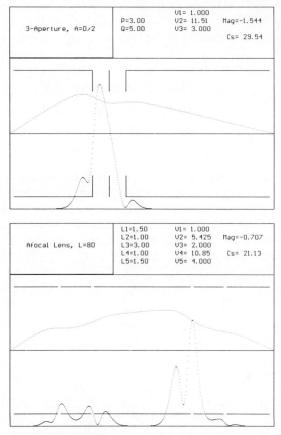

Figure 5.7 The integrands of the aberration integrals of equation (5.10) for three-aperture and Afocal8 lenses

maxima associated with each gap. The ray paths shown in this figure represent the true radius, $r(z)$.

The ratio of the reduced and true radii has the same value at any point for all trajectories, so we can write $R(z) = R_1(z) + (1/M)R_2(z)$ and expand the radial terms as

$$R^4 = R_1^4 + 4R_1^3 R_2/M + 6R_1^2 R_2^2/M^2 + 4R_1 R_2^3/M^3 + R_2^4/M^4$$

and similarly for $R^3 R'$. This allows us to separate the integral of equation (5.10) into the sum of five integrals, one for each of the C_{s_i}. Each integral is written as the product of terms involving the T-parameters

$$c_1 = \frac{1}{V^{1/2}} \left[3T^4 + 5T'^2 - \tfrac{11}{2} T^2 T' \right]$$

$$c_2 = \frac{4TT'}{V^{1/2}}$$

which, being functions of only the electrode geometry and potentials, are the same for all applications of the lens, and terms specific to each coefficient

$$I_0 = c_1 R_1^4 \qquad + c_2 R_1^3 R_1'$$
$$I_1 = c_1 4 R_1^3 R_2 + c_2 (3 R_1^2 R_2 R_1' + R_1^3 R_2')$$
$$I_2 = c_1 6 R_1^2 R_2^2 + c_2 (3 R_1 R_2^2 R_1' + 3 R_1^2 R_2 R_2') \qquad (5.11)$$
$$I_3 = c_1 4 R_1 R_2^3 + c_2 (R_2^3 R_1' + 3 R_1 R_2^2 R_2')$$
$$I_4 = c_1 R_2^4 \qquad + c_2 R_2^3 R_2'.$$

r_1 and r_2 are asymptotic to $\pm f_1$ which is not known *a priori* so the focal properties must be calculated using arbitrary horizontal asymptotes (usually 1) and then scaled to provide the values required for the aberration calculation. Figure 5.8† shows the integrands used to calculate the five C_{s_i}. The individual curves can easily be associated with particular coefficients and for this lens they are, from the top, $C_{s_2}, C_{s_0}, C_{s_4}, C_{s_3}, C_{s_1}$.

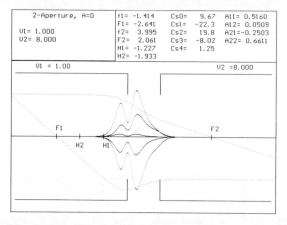

Figure 5.8   The integrands of the integrals analagous to equation (5.10) for the five spherical aberration coefficient integrals for a two-aperture lens. The curves can readily be associated with particular coefficients as these are markedly different in magnitude.

5.4   Interrelations of the aberration coefficients

By considering a lens to be 'thin', in the sense that R does not change over the distance for which T is significant, it is possible to factor a term out

---

† Figures 5.7 and 5.8 have been prepared with a specially modified version of **LENSYS**.

of the aberration integral and hence develop simple relationships between the m_{ij} coefficients [12]. It has been demonstrated [10] that these remain reasonably valid even when the 'thin' approximation does not appear to be appropriate. The relationships are of the form $m_{2j}/m_{1j} = \sigma$ where the value of the ratio has been determined empirically as $\sigma = (V_2/V_1)^{1/4}$ for a two-cylinder lens. Taken in conjunction with the well established relationships between three pairs of m_{ij} (equation (5.4)) and ignoring the added '1.5' terms, which is a reasonable approximation except for very strong lenses, there appears to be a pattern of interrelations between all the m_{ij} coefficients,

$$m_{i6} \simeq \sigma m_{i5}/3 \simeq \sigma^2 m_{i4}/3 \simeq \sigma^3 m_{i3} \qquad m_{2j} \simeq \sigma m_{1j}.$$

With these factors, equation (5.2) becomes

$$C_s(M) = -m_{13}f_2\left[1 + 4\frac{\sigma}{M}\frac{f_1}{f_2} + 6\left(\frac{\sigma}{M}\right)^2\left(\frac{f_1}{f_2}\right)^2\right.$$
$$\left. + 4\left(\frac{\sigma}{M}\right)^3\left(\frac{f_1}{f_2}\right)^3 + \left(\frac{\sigma}{M}\right)^4\left(\frac{f_1}{f_2}\right)^4\right]. \qquad (5.12)$$

If σ is indeed $(V_2/V_1)^{1/4}$, then $\sigma^2 = -f_2/f_1$ so $\sigma f_1/f_2 = -1/\sigma$ and, writing C_{s_0} for $-m_{13}f_2$ we can express the spherical aberration coefficient for a given magnification as

$$C_s(M) = C_{s_0}\left(1 - 1/\sigma M\right)^4. \qquad (5.13)$$

A careful study of more complex lenses shows that this is indeed rather approximate, but another result based on the same relationships does seem to have a wider validity. We define a function, Y, by

$$Y = C_{s_0}\left(\frac{-\sigma f_1}{f_2}\right)^4 + C_{s_1}\left(\frac{-\sigma f_1}{f_2}\right)^3 + C_{s_2}\left(\frac{-\sigma f_1}{f_2}\right)^2 + C_{s_3}\left(\frac{-\sigma f_1}{f_2}\right) + C_{s_4}$$
$$(5.14)$$

and substitute for the C_{s_i} the appropriate terms of equation (5.12), giving

$$Y = C_{s_0}\left(\frac{-\sigma f_1}{f_2}\right)^4 (1 - 4 + 6 - 4 + 1) = 0.$$

If we write Y_+ to represent the sum of the first, third and fifth terms of equation (5.14) and Y_- to represent the sum of the second and fourth (which is usually negative), then the ratio $(Y_+ + Y_-)/(Y_+ - Y_-)$ will be a fair measure of the accuracy of the statement $Y = 0$. For a wide range of lenses (and the reader is invited to test this using **LENSYS**) this ratio is less than 0.01 and usually much less.

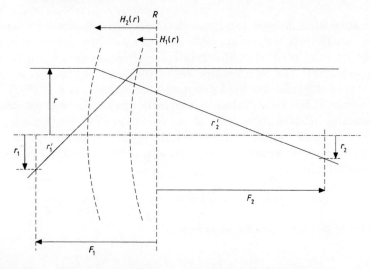

Figure 5.9   Diagram to illustrate the curved principal surfaces

5.4.1   The principal surfaces

We noted in Chapter 1 that the principal 'planes' were not in fact flat and we are now in a position to examine their true form. Figure 5.9 shows the principal surfaces as defined by the intersection of the object and image asymptotes of rays incident at a radial distance, r, parallel to the axis. To third order in r, we can express the distance from the second *Gaussian* focal point to the projection of the intersection on to the axis, $H_2(r)$, as

$$F_2 - H_2(r) = (r_2 - r)/r_2'$$

using equations (5.1)

$$F_2 - H_2(r) = f_2 + (f_1 m_{26} + f_2 m_{16}) \left(\frac{r}{f_2} \right)^2$$

and writing $\sigma^2 = -f_2/f_1$

$$F_2 - H_2(r) = f_2 \left[1 + m_{13}\sigma^2(\sigma - 1) \left(\frac{r}{f_2} \right)^2 \right].$$

A similar analysis gives, for the first principal surface,

$$F_1 - H_1(r) = (r_1 - r)/r_1'$$

$$= f_1 + (f_1 m_{23} + f_2 m_{13}) \left(\frac{r}{f_1} \right)^2$$

$$= f_1 \left[1 - m_{13}\sigma(\sigma - 1) \left(\frac{r}{f_1} \right)^2 \right].$$

Note that m_{13} is negative and $\sigma > 1$ for an accelerating lens, so both principal surfaces are concave towards the high potential side of the lens.

5.4.2 The aberrations of retarding lenses

Equations (5.1), which describe the ray position and slope at the second focal plane in terms of the values of these parameters at the first focal plane, may be inverted algebraically to describe the effect of reversing the direction of the rays

$$r_1' = -\frac{r_2}{f_1} + m_{26}r_2'^3 + m_{25}r_2'^2\frac{r_2}{f_1} + m_{24}r_2'\left(\frac{r_2}{f_1}\right)^2 + m_{23}\left(\frac{r_2}{f_1}\right)^3 + \cdots$$

$$\frac{r_1}{f_2} = -r_2' - m_{16}r_2'^3 - m_{15}r_2'^2\frac{r_2}{f_1} - m_{14}r_2'\left(\frac{r_2}{f_1}\right)^2 - m_{13}\left(\frac{r_2}{f_1}\right)^3 + \cdots$$

$$\text{(5.15)}$$

or equations (5.1) may be written in terms of the coefficients of the lens having the reciprocal voltage ratio

$$r_2' = -\frac{r_1}{f_2^*} + m_{13}^*r_1'^3 + m_{14}^*r_1'^2\frac{r_1}{f_2^*} + m_{15}^*r_1'\left(\frac{r_1}{f_2^*}\right)^2 + m_{16}^*\left(\frac{r_1}{f_2^*}\right)^3 + \cdots$$

$$-\frac{r_2}{f_1^*} = r_1' + m_{23}^*r_1'^3 + m_{24}^*r_1'^2\frac{r_1}{f_2^*} + m_{25}^*r_1'\left(\frac{r_1}{f_2^*}\right)^2 + m_{26}^*\left(\frac{r_1}{f_2^*}\right)^3 + \cdots$$

$$\text{(5.16)}$$

where the asterisks (*) denote the coefficients and focal lengths of the reversed lens. A term by term comparison of equations (5.15) and (5.16) shows that each coefficient for the reversed lens is the same as one for the forward lens, with an obvious symmetry

$$m_{13}^* = m_{26} \qquad m_{14}^* = m_{25} \qquad m_{15}^* = m_{24} \qquad m_{16}^* = m_{23}$$

and also $f_1^* = -f_2$. Notice that as we have not specified which lens is the accelerating one, this set of equations can be read either way.

Using equation (5.5) we can develop the relation between the coefficients of spherical aberration as

$$C_{s_0} = -m_{13}f_2$$

so

$$C_{s_0}^* = -m_{13}^*f_2^* = -m_{26}f_2^*$$

substituting $C_{s_4} = -m_{26}(f_1^4/f_2^3)$ gives

$$C_{s_0}^* = C_{s_4}f_2^*\left(f_2^3/f_1^4\right)$$
$$= -C_{s_4}\left(f_2/f_1\right)^3 \qquad \text{as } f_2^* = -f_1$$
$$= C_{s_4}\left(V_n/V_1\right)^{3/2}$$

where V_n is the potential of the last electrode and the ratio is that for the unasterisked direction. It is easy to show, by similar arguments, that

$C_{s_1}^* = C_{s_3}(V_n/V_1)^{3/2}$ and $C_{s_2}^* = C_{s_2}(V_n/V_1)^{3/2}$. Knowing that the reverse magnification $M^* = 1/M$ we can write equation (5.3) for the reversed direction as

$$C_s^*(M^*) = \left(\frac{V_n}{V_1}\right)^{3/2}[C_{s_4} + C_{s_3}/M^* + C_{s_2}/M^{*2} + C_{s_1}/M^{*3} + C_{s_0}/M^{*4}]$$

$$= \left(\frac{V_n}{V_1}\right)^{3/2}[C_{s_4} + MC_{s_3} + M^2C_{s_2} + M^3C_{s_1} + M^4C_{s_0}]$$

$$= M^4\left(\frac{V_n}{V_1}\right)^{3/2}[C_{s_4}/M^4 + C_{s_3}/M^3 + C_{s_2}/M^2 + C_{s_1}/M + C_{s_0}]$$

$$= M^4\left(\frac{V_n}{V_1}\right)^{3/2}C_s(M). \tag{5.17}$$

In the case of the afocal five-cylinder lens discussed in Chapter 4, the magnification is given simply by $M_{af} = (V_5/V_1)^{-1/4}$, so the ratio of the spherical aberration coefficients in the forward and reverse directions is $(V_5/V_1)^{1/2}$.

5.4.3   Figures of merit

The geometric aberrations of a lens system for a given application may be reduced by the correct choice of the number and structure of the electrodes. The application will usually specify the potential ratio of the image and object spaces, the separation of the object and image, the magnification to be used and the maximum acceptible pencil angle of the beam in one or other of the spaces. The desirable parameter will be a minimum size of aberration disc. The diameter of the lens system might be restricted, but it is frequently a free parameter. We can write the radius of the aberration disc as

$$\frac{\Delta r}{D} = MC_s(M)r_1'^3$$

where we express the coefficient in units of D. The value of the diameter also appears in the actual length, L, of the system in terms of lens parameters, as

$$\frac{L}{D} = \frac{Q-P}{D} = F_2 + q - F_1 - p = F_2 - F_1 - Mf_2 + f_1/M \tag{5.18}$$

and so we can write

$$\Delta r = \frac{LMC_s(M)}{F_2 - F_1 - Mf_2 + f_1/M}r_1'^3. \tag{5.19}$$

It would plainly be difficult to examine all possible lenses to find that which gave the smallest value of Δr, but, by making two reasonable approximations, it is easy to find a good criterion to point us towards a suitable lens.

The first approximation is to use equation (5.13) to express the spherical aberration coefficient in terms of the magnification and overall voltage ratio. The second is to equate the arithmetic mean of the mid-focal distances to the geometric mean of the focal lengths.† The denominator of equation (5.19) becomes

$$2(-f_1 f_2)^{1/2} - M f_2 + f_1/M = - M f_2 (1 - 1/\sigma M)^2$$

and so

$$\Delta r = - L C_{s_0} (1 - 1/\sigma M)^2 r_1'^3 / f_2$$

which can be separated into

$$\Delta r = - g L (1 - 1/\sigma M)^2 r_1'^3$$
$$g = C_{s_0}/f_2 = -m_{13}$$

where g contains all the lens-dependent information and none of the specification and is therefore a suitable figure of merit for the inter-comparison of lenses. For the type of problem considered here, where the overall voltage ratio forms part of the specification, it will almost certainly be the case that a lens of three or more elements will have an appreciably smaller value of g than any two-element lens. Once the lens geometry and inner voltage ratios have been selected, the choice of lens diameter follows from equation (5.18) where the focal properties are in units of D.

For the case of rays focused from infinity, for which the magnification is zero, this analysis fails because the distance between conjugates is infinite. However, the appropriate specification in this case is more likely to be in terms of the distance of the image from the lens itself, probably specified as $L = F_2$, the mid-focal distance. The radius of the aberration disc is given by equation (5.6) and the figure of merit is just $C_{s_4}/f_1^3 = -m_{26}(f_1/f_2)$, though, as the ratio of focal lengths is implicit in the specification of the voltage ratio, $-m_{26}$ is itself suitable.

If the specification allows for an adjustable magnification, the quantity to be optimised becomes the product $M C_s$ which, again using equation (5.13), can be written as

$$M C_{s(M)} = M C_{s_0} (1 - 1/\sigma M)^4.$$

---

† This result can be shown to be true for lenses weak enough that $Y = 0$, where Y is defined by equation (5.14).

We may differentiate this to find an optimum value of $M = -3/\sigma$, which gives for the minimum radius spot in the Gaussian image plane

$$\Delta r = -9.48(C_{s_0}/\sigma)r_1'^3 = 2.37C_{s_1}r_1'^3$$

and suggests that m_{23} would be a better figure of merit with this looser constraint.

Chapter 6

The LENSYS program

6.1 The programs on the disc

The disc which accompanies this book contains three files for use on an IBM or compatible personal computer. The file **LENSYS.EXE** is the program file which provides facilities for a broad range of studies on round electrostatic lenses. The file **POTS.DAT**, which is loaded automatically by **LENSYS**, contains data from which the potentials and potential gradients are calculated. Both files should be copied to another floppy disc for use and the original put away in a safe place. The files may be copied on to a hard disc if desired. The third file, **RELAX51F.EXE**, is included to allow the reader to examine the convergence of the iterative procedure discussed in Chapter 3 and illustrated in figure 3.3.

The program will not run in a machine with only 256k of memory: 384k is probably enough, depending on the presence of other software such as **SIDEKICK** or graphics drivers for **SIMION**. If the program finds that there is insufficient memory, it will stop with a message saying that there has been a memory overflow error.

A graphics screen is necessary. VGA is best, but EGA is acceptable and, though the program will use a CGA screen, the resolution is only just adequate, especially for text messages and data tables. The Hercules monochrome graphics screen is usually very clear, but the lack of colour is not a trivial loss. The one place where the program behaves differently on a monochrome screen is in the menus, in which the selected item appears in inverse video rather than in a different colour.

6.1.1 Loading and running the programs

The two files should both be in the current directory as **LENSYS** will not look elsewhere for the data file. The program is started by typing 'LENSYS' at the DOS prompt and it presents a screen with brief details of the program. If it does not find a maths coprocessor in the computer, it writes a message to the screen telling you that it would run much faster with one, but it continues nevertheless. Once this introductory screen has been displayed, it reads the data file and will then make no further use of the disc, either to read or write. If you are running the program from a floppy disc, it may be removed at this point. After the data file has been loaded, you are invited to 'Press ENTER to continue'. This starts the program and shows a screen which will provide the communication channel.

6.2   The screen display

The lower three quarters of the screen is the graphics window. This will show schematic representations of the lens being studied and the rays traced during the analysis of the lens. The breadth of this window is equivalent to 8 lens diameters. The upper portion of the screen forms two text windows. The wider one, to the right, is called the data window and is for the display of lens data and the results of the calculations, while that to the left is called the message window and shows the menus, together with other prompts and messages in the course of the program. It is usually of a different colour from the data window. The manual input of lens parameters (potentials, lengths, etc) is normally made to the data window, but sometimes to the message window.

6.2.1   The menu system

All choices are made via a system of menus, each offering between two and five options. The opening menu describes itself as 'Main Menu' and invites a choice between

Lens Data

Imaging

Exit to DOS

with the top entry displayed in a different colour (inverse video on a monochrome screen) to signify that it is the 'selected' item. The selection may be changed by the UP- and DOWN-arrow keys and, as the motion rolls over, the quickest way to select the bottom item is to use the UP key once. Pressing the ENTER key makes the choice. Choosing 'Exit to DOS' is the normal exit route from the program. If it is chosen, the program does not request verification, but exits immediately with a simple closing message leaving the display in a 25 line text mode. The program does not remain in memory and has to be reloaded if required again.

6.3   The Lens Data option

Choosing 'Lens Data' takes you to another menu 'Select a Lens' which offers the choice of

2-Aperture A=D/2
2-Aperture A=D
3-Aperture A=D/2
2-Cylinder G=D/10
3-Cylinder G=D/10.

This section allows you to find the focal properties of a lens. The geometries of the first four lenses are predefined, but if you choose the three-cylinder lens you are asked, in the message window, for the length of the second element, in units of the diameter, and this should not exceed 3. If you enter too large a value, the invitation is repeated. No sign is expected, only two places of decimals will be accepted and the length will be rounded down to the next multiple of 0.05. The program then draws a schematic of the lens in the graphics window and asks for the voltage ratios, V_3/V_1 (if appropriate) and V_2/V_1. These are entered in the message window. Up to three decimal places are accepted here. All voltage values presented in the data window are on the basis that $V_1 = 1$ and so V_n represents V_n/V_1. The chosen values of the electrode potentials are written in the graphics window, above the schematic electrodes. The message window will tell you that the program is calculating the trajectories and the aberrations of the lens and two rays will be drawn in the graphics window. These are the usual principal rays, drawn from the left and from the right, parallel to the axis. The focal points are marked and labelled as are the principal points. The data window shows three columns of data. The first column displays values of the focal lengths, mid-focal distances and the positions of the principal points, the second column the values of the five spherical aberration coefficients, C_{s_i}, and the last column the values of the four elements of the bending matrix described in Chapter 1.

When the calculation and display are complete, the message window shows a further menu headed 'Continue?'.

New Potentials
New Lens
Print
Main Menu

We shall discuss 'Print' later. The 'New Lens' option leaves you in the 'Lens Data' part of the program by returning you to 'Select a Lens'. If you wish to examine the properties of a lens between conjugates it is necessary to choose 'Main Menu' and then 'Imaging' from that menu.

6.3.1 Error conditions

There are a number of input parameter values which the program will not reject immediately, but it will discover that they are not sensible and will halt with an appropriate message and wait for ENTER to be pressed. A very simple error is to input $V_2 = 1$ for a two-element lens which gives no lens action at all. Another input leading to error conditions is an excessively high voltage ratio. There are a number of tests for these. If the emergent ray has the wrong *sign* of slope it has probably crossed the axis twice. If the ray crosses the axis between the outermost electrodes, the focal length will not be a very useful parameter and it would be misleading to display the results in the data window.

6.4 The Imaging option

This opens with a further menu, 'Select Lens Type' and you have to choose between

Aperture Lenses

Cylinder Lenses.

The details of the lens specifications and the positions of the conjugate points have to be entered in a slightly different fashion for the two lens types. These differences will be explained later. The program traces a ray starting from the axial object point at the left of the graphics window and with a slope of 1, but the subsequent action depends on a further choice which you will have been asked to make by a menu headed 'Adjust Potentials'

Automatically

Manually.

6.4.1 Automatic and manual adjustments

If you choose automatic adjustment, the potential of one electrode will be adjusted, in a series of steps, until the ray crosses the axis at the image point you have specified. Each ray is deleted as the next ray is being drawn. In the case of lenses with two or three elements the potential of the second electrode is adjusted, but if there are more than three electrodes you are asked, by a message, to indicate which one. You may not choose the first or the last electrode. After the program has made an automatic adjustment it draws two further rays. These both start from an object point above the axis and one has a slope of zero and the other a slope of -1. These rays will cross at the image plane and indicate the magnification. The data window shows the lens specification (a little differently in the two cases), the electrode potentials, updated at each iteration step, and the magnification and spherical aberration coefficient.

For any lens geometry, more than one set of electrode potentials will satisfy the conjugate focus condition. It is possible to persuade the program to pick one combination rather than another by choosing a suitable starting point for the potential which is to be varied. For example, even a two-element lens has two possible voltage ratios, one accelerating and the other decelerating, and if V_2 is initially set to 0.1 the program should find the retarding solution. However, if this requires a potential of less than 0.01 it will fail, with a message 'This is unlikely to converge. ENTER continues'. A three-element lens started with $V_3 = V_2 = V_1 = 1$ will converge to the solution with $V_2 > 1$. There is a general tendency built into the program to look for the higher potential solutions because these are usually the ones with smaller aberrations.

With manual adjustment, only a single ray is drawn and the 'miss distances', ΔR and Δr, at the image plane are displayed. You are then asked if you wish to draw another ray. If so, you are asked for the starting position and slope of the ray and it is then drawn and the 'Another Ray?' question repeated. None of these rays are deleted until you exit by declining the invitation. This mode allows a particular set of rays to be drawn for a lens for which the automatic adjustment has already provided values of the electrode potentials. It might, for example, be used to find the position of the image of a second aperture, an entrance pupil say, which can be simulated by the intersection of a ray parallel to the axis and some distance above it with the original ray. It can also provide a way to study a lens which the program has rejected as being too strong or not useful. Neither the magnification nor the spherical aberration coefficient is calculated in this mode.

At the end of the automatic mode or after exiting from the manual mode, the message window shows the same 'Continue?' menu as in the Lens Data case, but now the choice of 'New Lens' returns you to the 'Select Lens Type' menu.

6.4.1.1 Error conditions.   If the automatic routine cannot find a value which allows the ray to pass through the axial image point, it will issue a message to that effect and wait for ENTER to be pressed. This will sometimes happen because it is looking in the wrong region. For example, it might be trying very small values and meeting the $\not< 0.01$ constraint while a much larger value may be valid. This condition can usually be caught by noting the way in which the variable potential changes after each iteration. A more common cause for failure is that the overall voltage ratio is too high. If the potential of the variable electrode is observed to be varying around, or near to, the geometric mean of the neighbouring potentials, this can be taken as a sign that that region of the lens is too strong and new fixed potentials should be tried. It is often worth while to try some values close to these critical ones, using the Manual option, to assess the nature of the change that is required.

6.4.2 Aperture lenses

A menu offers you the choice of the three aperture lenses already seen in the Lens Data option. A message asks for values of the object and image distances, P and Q, with a constraint that the total distance from object to image should not exceed 8 diameters. Despite the sign convention requiring P to be negative, only the magnitude should be entered as the program can only handle real objects and images and there can, therefore, be no ambiguity. The values appear in the data window. A schematic diagram of the lens is drawn in the graphics window, with the object at the left hand edge of the window, and the centre of the lens a distance P to the right. A vertical line is drawn across the axis a further distance Q beyond the lens centre. You are then asked for the electrode potentials which are written in the data window replacing '?' prompts. Next the 'Adjust Potentials' menu appears and the subsequent behaviour is as we have already discussed.

6.4.3 Cylinder lenses

A menu, 'Select a Lens', offers a choice of various cylinder lenses

2–5 Cylinder
Afocal6
Afocal8
Varimag6
Varimag8.

The last four of these will be discussed later. Choosing the first option leads to yet another menu, 'How Many Cylinders', and thence to the particular multi-cylinder lens you require.

In contrast to the aperture lenses for which only a limited set of spacings are available in this program, the cylinder lenses may use a range of lengths of cylinders and it is rather simpler to specify the object and image positions in terms of these lengths. This is particularly the case for a lens with cylinder lengths of (say) 2, 1, 2 and 3 diameters where the choice of reference plane might be a matter for argument! The program, therefore, expects the object to be at the left hand end of the first cylinder and the image at the right hand end of the last cylinder, though it will extend the last cylinder to the right hand edge of the screen in the schematic diagram, marking the image plane with a transverse line as before. Having chosen the number of cylinders, you are asked for their lengths (which will be rounded down to the next multiple of $D/20$) and their potentials, which may be specified to three decimal places. The values of the lengths are taken to include a half share of the inter-cylinder gap of $D/10$ and the total length must not exceed 8 diameters. As a trivial example, a lens with $L_1 = 2.00, L_2 = 1.00$ and $L_3 = 3.00$ (for which the reference plane is unambiguously in the middle of the second element) would have P= 2.50 and Q= 3.50. The 'Adjust Potentials' menu then appears and the subsequent behaviour is as we have already discussed. Lenses with three cylinders are widely used to satisfy the simple condition of maintaining a conjugate pair of points for a range of overall voltage ratios. An example was shown in figure 4.6. The length of the centre element affects the behaviour of both the magnification and the spherical aberration, though not to a great degree. The addition of a fourth cylinder allows for better control of the magnification, but a fifth cylinder opens the possibility of controlling not just the magnification, but also the spherical aberration of the lens system.

6.4.3.1   The aberration behaviour of the five-cylinder lens.   The conjugate focus condition for a lens having the potentials of the third and fifth electrodes fixed may be achieved with each of the intermediate potentials at a high or low value compared with their neighbours. The four possible combinations lead to significantly different coefficients of spherical aberration for all values of the magnification. This is illustrated in figure 6.1 for a five-cylinder lens having $V_5 = V_3 = V_1 = 1$. For all values of the magnification, the spherical aberration coefficient is least for $V_2/V_1 > 1$ and $V_4/V_3 > 1$.

6.4.4   The Afocal and Varimag lenses

These lenses each have five cylinder elements with lengths

	for the "6"	and the "8" lenses
$L_1 =$	1.25	= 1.50
$L_2 =$	0.50	= 1.00
$L_3 =$	2.50	= 3.00
$L_4 =$	0.50	= 1.00
$L_5 =$	1.25	= 1.50.

6.4.4.1   The Afocal lens.   The afocal lens was discussed in Chapter 4 and the only input required is the value of the overall voltage ratio. The program automatically sets $V_3 = V_5^{1/2}$ and adjusts the other potentials maintaining $V_4/V_3 = V_2/V_1$. The data window reflects the changes in the potentials and, after convergence, shows the magnification and the spherical

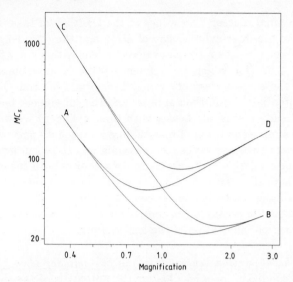

Figure 6.1   The spherical aberration, expressed as the product MC_s, of a five-cylinder lens with $V_5 = V_3 = V_1$ as a function of the magnification for values of V_2 and V_4 in the following ranges:

Region	V_2	V_4
AB	> 1	> 1
BC	> 1	< 1
CD	< 1	< 1
DA	< 1	> 1

aberration coefficient. The program operates only in the regime with V_2 and V_4 both at the higher of their two possible values. The shorter lens has much smaller aberrations, but the end elements are rather short. Placing a real angle stop within the electrode would therefore affect the potential distribution to a certain extent.

6.4.4.2   The Varimag lens.   These are very powerful lenses which allow the magnification and the overall voltage ratio to be specified and a further parameter to be adjusted. This may be either the magnification of an entrance pupil or the spherical aberration coefficient of the lens. These are controlled, though not independently, by the potential of the centre electrode. The program assumes that there is an entrance pupil at a distance $D/2$ from the entrance window. You are asked to supply values for V_5, V_3 and the target magnification. This will be a negative quantity, but your input will be read and displayed in the initial data as the modulus, |MagT|. The program first examines the range of magnifications which can be achieved with these potential values, and may halt with a message to the effect that the target magnification is outside the usable range. Otherwise it adjusts V_2 to satisfy the conjugate condition and then adjusts V_4 to change the magnification. Both V_2 and V_4 are constrained to take the higher of their possible values.

These adjustments of V_2 and V_4 continue alternately until the window magnification is close enough to the target value. At this point the data window, which has monitored the potentials and the window magnification throughout, displays the target, window, pupil and longitudinal magnifications (MagT, MagW, MagP & MagL) and the spherical aberration coefficient of the lens. A menu invites a choice of

Modify V3

Continue.

The first allows you to change V_3 and the whole convergence procedure is repeated while the second opens the usual 'Continue?' menu. Increasing V_3 from the 'afocal' value of $V_5^{1/2}$ will both decrease the pupil magnification and reduce the spherical aberration. If the target magnification is close to one of the limiting values with the initial value of V_3, it is possible that changing V_3 might modify the limit sufficiently to cause the program to fail to converge. If this happens, it will be necessary either to change the target magnification or to accept a restriction on the value of V_3.

In the case of the 6 diameter lens, the exit pupil can be seen, provided that the longitudinal magnification is less than 4. The reader will notice, if this lens is operated with $V_5 = V_3 = 1$ and MagT$= -1$, that the pupil magnification is not quite equal to -1. This is because the potential 1.25 diameters inside a cylinder differs sufficiently from the potential of the electrode itself such that the trajectories are not quite straight. The effect is significant only at the 1% level and, for the 8 diameter lens, it is even smaller.

6.5   The print option

A simple pixel level dump of the screen can be made to a suitable printer connected to the first parallel port (LPT1) of a personal computer. On selecting Print from the 'Continue?' menu, a second menu, 'Which Region?' invites a choice of

Whole Screen

Data Window

Graphics Window

and a third menu offers

Epson

Laser.

The message window then clears and some information about the lens is displayed there. The printing is done by reading each pixel and copying it as a black dot unless it has the colour of the background. The DOS Graphics command is not used and the quality of the printed output reflects that of the monitor, but with all the colours lost. Figures 2.1(*b*), 5.7 and 5.8 were printed from a VGA monitor and figure 6.2 shows that from a CGA and from an EGA monitor. The number of dots per inch used by the different printers and the different proportions of the various monitor screens expressed in pixels imply a range of size and aspect ratio in the printed output. The program issues two Line Feed characters after printing, but no Form Feed character. A VGA display printed to a laser printer retains the correct proportions and is of such a size that two complete screens or seven data windows may be printed on A4 paper.

If the printer is not connected, not turned on or has run out of paper a message will be issued and the program will wait for ENTER to be pressed, at which point it will return to the 'Continue?' menu.

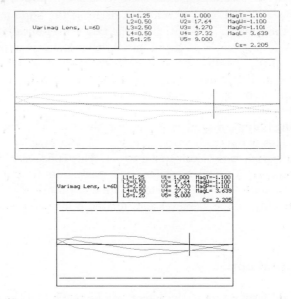

Figure 6.2   Printed output from EGA (upper) and CGA (lower) screens of a Varimag6 display

Appendix

Technical aspects of LENSYS

A.1   The programming language

The **LENSYS.EXE** program is written mainly in Turbo Pascal with some sections in Assembly Language. The floating point variables are of type double and require a maths coprocessor or its emulation: this latter is included in the code and will be used automatically if needed.

A.2   Initialisation

The program first looks for a maths coprocessor and, if it does not find one, writes a message to the opening screen. It then loads the data file, stores the contents in dynamic memory and waits for a response from the user. The type of Graphics adaptor is then examined and various scale and colour parameters are set in the light of this. If a CGA adaptor is found, it is set to the four-colour resolution of 320 by 200. This does not affect the effective resolution of the display seriously and is preferred to the two-colour 640 by 200 mode.

A.2.1   The effect of processor type on the speed of the program

There are three interlinked factors which affect the speed at which the **LENSYS** program runs. The 8088, 80286 and 80386 processors are inherently faster in that order, and the clock speeds of the 80286 and 80386 processors are normally much greater than that used for the 8088, even in 'turbo' versions. A maths coprocessor has a very marked effect, especially with the 8088. The table shows approximate timings for the 2-Aperture lens focal data routine with a voltage ratio of $10:1$. The values are the time interval between pressing ENTER and the completion of the display in the data window. Other parts of the program show the same ratios of speed, though the imaging routines naturally take more time. As an example, the Varimag6 routine, started with $V_5 = V_3 = 1$ and with a target magnification of 1 also, takes some eight times as long. This is probably the slowest of the routines, at least with reasonable values of the starting parameters.

Processor	Clock	Time
8088	8Mhz	80s
8088/87	8Mhz	5.0s
80286	12Mhz	18s
80286/287	12Mhz	3.5s
80386	25Mhz	6s
80386/387	25Mhz	1.3s

A.3   Calculation of potentials

The file **POTS.DAT** contains values of the axial potential and the potential gradient for all the lens geometries at intervals of $D/80$, which is the step size used by the ray tracing routines. A single exponential term is used to represent the potential and the gradient for $|z/D| > 3$ and there would have been little error in using this approximation for even smaller values of z/D. The following sections describe the methods used to calculate these values.

A.3.1   Aperture lenses

The potential distributions for the aperture lenses were calculated by the five-point relaxation method of equation (3.8) using lattices of 61 by 901 and 121 by 561 points with an aperture radius equivalent to 20 and 40 points, respectively. An acceleration parameter of 1.90 to 1.95 was used and the iteration continued for some 1000 steps. The criterion for convergence was that the sum of the moduli of the change in the potential at all axial points should be less than 10^{-6} for a potential difference between the electrodes of 1. The long thin lattice gave a good indication of the extent to which the end plate of the electrode affected the result and was also a guide to the axial distance beyond which the potential could be represented by a single exponential. The wider lattice matched the integration step used in the ray tracing routines. The axial potential gradient was calculated from a cubic spline fit to the potential.

A.3.2   Cylinder lenses

The potential inside two closely spaced $(D/10)$ cylinders is very well represented by an expansion in terms of Bessel functions and the particular method described in section 3.4 was used, including the correction to the effective length of the gap. The potential was checked using the extension to the relaxation method illustrated in figure 3.5, with the calculations extending to a lattice of 81 by 561 points. The Bessel function expansion also gives the potential gradient at any point with no need for curve fitting or averaging.

An outline Pascal program is shown on the following page. All details of input and output have been omitted as they are quite straightforward, yet might differ somewhat between specific applications. We set the electrode potentials to be $V_1 = 0$ and $V_2 = 1$. The input variables are $k[n]$, the nth zero of the Bessel function of order zero, and $j[n]$, the corresponding value of the Bessel function of order one. For convenience we define $q[n] = 1/(2 * j[n] * k[n] * k[n])$ and $r[n] = 2 * k[n] * q[n]$ as factors used in the evaluation of the potential and the gradient, respectively.

At each axial position the potential, V, and its gradient, E, are calculated by summing terms $dV[c]$ and $dE[c]$, respectively, until the gradient term is less than some limiting value (taken here as 10^{-7}) or the number of terms exceeds 150, when a rounding-off procedure is applied to estimate the sum to infinity. The expressions for the potential and the gradient all involve terms in $(\pm 2z \pm g)$ which we express as $z1 = 2z + g$ and $z2 = 2z - g$. We then define $a1$ and $a2$ so that, for $|z/D| > g/2$, both are less than one.

The series converge rather slowly close to the boundaries between the regions. We avoid the slowest convergence by arranging not to make the calculation exactly at the boundaries (which lie at $z/D = \pm 0.054\,125$), but even at $|z/D| = 0.0500$ and 0.0625 they have not converged adequately after 150 terms, and we need to invoke the round-off procedure. By considering the continuity of the gradient at the boundary between regions I and II, it is easy to show that

$$\sum_{n=1}^{\infty} \frac{1}{K_n J_n} \equiv \sum_{n=1}^{\infty} \frac{1}{k[n]j[n]} = 0.5.$$

The sum to 150 terms is, at $0.471\,15\ldots$, significantly different from this. The sums to odd and even numbers of terms differ from the true sum by amounts which alternate in sign and decrease progressively. A simple average introduces some bias as a result of this progressive decrease, but a mean of three successive terms is really very good. The round-off procedure used in this program is much more powerful than is really necessary and could probably be applied safely to estimate the sum to infinity after only the first 30 or so terms had been calculated. We use it to extrapolate from the sums to 146 to 150 terms by using

$$s[c+3] = \frac{s[c] \times s[c-2] - s^2[c-1]}{s[c] - 2s[c-1] + s[c-2]}$$

with $c=148$, 149 and 150 to generate pseudosums, $s[151]$, $s[152]$ and $s[153]$ which are then averaged again to give a final pseudosum, $s[156]$, which is

```
PROGRAM CylinderLensPotential;

CONST radius=40;                                { In units of step size }
      length=320;                               { Two cylinders 4D long }
      gap=0.1;                                   { With a gap of D/10 }

VAR   j,k,q,r:ARRAY [1..150] OF double;         { Bessel functions }
      V,E:ARRAY [-length..length] OF double;    { Potential and Gradient }
      sV,sE:ARRAY [0..156] OF double;           { Sums to 'c' terms }

PROCEDURE ReadData;                             { Modify to read from a file }
VAR n:integer;
BEGIN FOR n:=1 TO 150 DO BEGIN
  read(k[n],j[n]); q[n]:=-1/(2*j[n]*k[n]*k[n]); r[n]:=-2*k[n]*q[n]; END; END;

PROCEDURE RoundOff(c:integer);                  { Used to force convergence }
BEGIN sV[c+3]:=((sV[c]*sV[c-2]-sqr(sV[c-1]))/(sV[c]-2*sV[c-1]+sV[c-2]));
  sE[c+3]:=((sE[c]*sE[c-2]-sqr(sE[c-1]))/(sE[c]-2*sE[c-1]+sE[c-2]));    END;

PROCEDURE CalculatePotential;
VAR   l,c:integer;
      g,h,z,z1,z2,V0,E0,a1,a2,limit:double;{ Cylinder potentials,V0=0,1 }
      dV,dE:ARRAY [0..150] OF double;          { Increments in V and E }

BEGIN
  limit:=0.0000001;  h:=1/(2*radius);
  g:=1.0825*gap;                           { Allows for non-uniform field in gap }
  FOR l:=-length TO length DO
  BEGIN
    z:=l*h;      z1:=2*z+g;      z2:=2*z-g;
    IF z > g/2 THEN BEGIN    V0:=1;      E0:=0;      END
                ELSE IF z < -g/2 THEN BEGIN    V0:=0;   E0:=0;         END
                            ELSE BEGIN V0:=0.5 + z/g; E0:=-1/g;    END;
    c:=1;    sV[0]:=0;   sE[0]:=0;   dE[0]:=1; { Or it will not start }
    WHILE   s(dE[c-1]) > limit DO
    BEGIN
      IF z > 0 THEN BEGIN  a1:=exp(-k[c]*z1); a2:=exp(-k[c]*z2);   END
                ELSE BEGIN  a1:=exp(k[c]*z1);  a2:=exp(k[c]*z2);    END;
      IF    s(z) > g/2 THEN BEGIN    dV[c]:=q[c]*(a1-a2)/g;
                                     dE[c]:=dV[c]*2*k[c]*(z/  s(z));       END
{z=0 to g/2}          ELSE IF z > 0 THEN BEGIN  dV[c]:=q[c]*(a1-1/a2)/g;
                                                dE[c]:=r[c]*(a1+1/a2)/g;  END
{z=-g/2 to 0}          ELSE BEGIN  dV[c]:=q[c]*(1/a1 - a2)/g;
                                   dE[c]:=r[c]*(1/a1 + a2)/g;    END;
      sV[c]:=sV[c-1]+dV[c];      V[l]:=V0+sV[c];
      sE[c]:=sE[c-1]+dE[c];      E[l]:=E0+sE[c];
      IF c < 150 THEN c:=c+1  ELSE BEGIN    RoundOff(148);  { Not converged }
                                            RoundOff(149);
                                            RoundOff(150);
                                            RoundOff(153);
                                            V[l]:=V0+sV[156];
                                            E[l]:=E0+sE[156];
                                            dE[149]:=0;        END;
    END;
  END;
END;

BEGIN   ReadData; CalculatePotential; END. { Modify to write to a file }
```

within 10^{-10} of 0.50 for the exact boundary case. Away from the boundaries, the series converge quite quickly: 20 terms being enough by $|z/D| = 0.15$.

To determine the potential distribution in a multi-cylinder lens, we write

$$V_{i,i+1}[l] = V[l + l_{i,i+1}] \qquad \text{and} \qquad E_{i,i+1}[l] = E[l + l_{i,i+1}]$$

where $l_{i,i+1}$ represents the position of the gap between the ith and $(i+1)^{\text{th}}$ cylinders. We can then represent the axial potential and its gradient in terms of the electrode potentials as

$$VA[l] = V_1 + (V_2 - V_1)V_{12}[l] \qquad EA[l] = (V_2 - V_1)E_{12}[l]$$
$$+ (V_3 - V_2)V_{23}[l] \qquad + (V_3 - V_2)E_{23}[l]$$
$$+ (V_4 - V_3)V_{34}[l] \qquad + (V_4 - V_3)E_{34}[l]$$
$$+ (V_5 - V_4)V_{45}[l] \qquad + (V_5 - V_4)E_{45}[l]$$

and then calculate

$$T[l] = EA[l]/VA[l]$$
$$T2[l] = 0.1875 * T^2[l]$$

where $T2 \equiv T^*$ of Chapter 4, and

$$F[l] = VA^{1/4}[l]$$

which is needed for conversion from real to reduced radii and *vice versa*.

A.4   Ray tracing for focal data

All the ray tracing is done by integrating the Picht equation using Numerov's algorithm (equation (4.3)). This requires two initial values of the reduced radius to be known. These are taken to be $RF[0] = F[0]$ and $RF[1] = F[1]$ for the forward direction, and $RB[639] = F[639]$ and $RB[638] = F[638]$ for the reverse direction. Both pairs of values correspond to a real radius of one. Subsequent values of $RF[l]$ are calculated and, if $l \bmod 4 = 0$, a point is plotted to the screen at $(l, RF[l]/F[l])$ with the ordinate suitably scaled to take account of the different resolutions of the various graphics cards. If $RF[l+1]/RF[l] < 0$ then the ray has crossed the axis and the position of the first focal point is found by interpolation. The position of the corresponding principal point is calculated from the slope of the *real* ray, extrapolating back to a radius of 1. The focal length is then found by subtraction. If the ray has not crossed the axis by the end of the calculation (i.e. if $RF[639]$ is still positive), which will happen for weak lenses, the focal properties are all found by extrapolation. Values for the reverse direction are determined in exactly the same manner. The matrix elements are also calculated.

The coefficients of spherical aberration are calculated using equation (5.10) with the integrands expressed by equation (5.11). $RF[l]$ and $RB[l]$ are first scaled to give $R1[l] = f_1 * RB[l]$ and $R2[l] = -f_1 * RF[l]$, and the integrals are evaluated using Simpson's rule.

There are certain conditions which will halt the program. If the slope of the ray at the exit plane is zero, and if the ray has not crossed the axis, then there is no lens action. If the ray crosses the axis between the lens centre and the outer electrode, it is treated as being too strong to be useful. If the slope is positive at the exit plane, then the ray must have crossed, or be about to cross, the axis a second time, and the lens is unlikely to be useful.

A.4.1   Accuracy and consistency

The lens data are presented to four significant figures, though the fourth should be ignored. The calculation of the forward and reverse characteristics are made independently and the ratio of the focal lengths provides a test of the precision of the ray tracing routines. This ratio should equal the square root of the overall voltage ratio. The determinant of the lens matrix should equal $-f_1/f_2$. Further tests may be made by examining lenses which are reversals of each other, for example 10:1 and 1:10, or 1:8:2 and 1:4:0.5 for a three-element lens. These tests can also examine the coefficients of spherical aberration, which should be related by equation (5.17).

A.5   Solving the imaging problem

The two initial values of the reduced radius are taken to be $R[0] = 0$ and $R[1] = h * F[1]$, corresponding to a slope of one for the real radius. The Picht equation is again integrated step by step for the whole length of the system, and each fourth point is plotted, again as real radius. The value of $R[I]$ is noted, where I represents the image plane. In the manual adjustment mode, this value, and $r[I]$, the real radius at the image plane, are displayed. In the automatic adjustment mode, the focusing voltage, V_f, is adjusted by an amount which depends on this miss distance as

$$V_f := V_f + k \times R[I].$$

After the first trace, k can only be guessed and a value of 0.1 is taken in order that the effect will be quite small. After this first correction, a suitable value for k can be estimated, because $R[I]$ is then known for two values of V_f, and we write

$$k = (V_f - V_f^\dagger)/(R^\dagger[I] - R[I])$$

where the $\dagger$ indicates the *previous* value.

The first change is deliberately made small, because even the sign of k is not initially known. If $R[I]$ is positive, the lens is too weak and must be strengthened. The correct value of V_f may have one of two values: 'high' or 'low', with the 'high' value generally offering better aberration behaviour. However, there are occasions when the 'low' value is of interest. In this case, the appropriate change in V_f is a reduction. The small initial change is a way of letting the program know where the target value of V_f lies.

If the change in V_f leads to a value less than 0.01, it is replaced by a small random value, greater than 0.01. The value is small to keep the search for V_f in the 'low' region, and random so that the program does not immediately get stuck in a loop. The target for satisfactory imaging is that $|R[I]| \leq 0.001$.

When the correct value of V_f has been found, the coefficient of spherical aberration is calculated using equation (5.10). Two further rays are then drawn from a point above the axis; one is initially parallel to the axis, and the other has a slope of -1 in real space. These two rays cross again at the image plane, and the magnification can be calculated from the final and initial off-axis positions. The exact starting point is chosen to be such that the magnified image lies within the length of the marker drawn perpendicular to the axis at the image plane. This requires that the program knows the magnification before these rays are drawn. The reader should have no difficulty in working out how this is possible!

A.5.1 The Afocal lens

This is handled in exactly the same way as any other lens. Though only V_5 is specified by the user, $V_3 = V_5^{1/2}$ and the *two* focusing voltages are related by $V_4 = V_2 \times V_3$. Only the higher values of the focusing voltages are considered. This is achieved by setting the lower limit for V_2 not at 0.01, but at $\sqrt{V_3}$, a value for which the $V_1 - V_2 - V_3$ lens is close to minimum power†.

A.5.2 The Varimag lens

Constraints are first applied to the possible values of $V_2 (\geq V_3^{1/2})$ and $V_4 (\geq (V_3 \times V_5)^{1/2})$. V_2 is set to its minimum value, and V_4 is first set to the greater

---

† The reader might like to examine the focal properties of three-element lenses with the centre element at potentials between those of the outer two.

of V_3 and V_5, and then adjusted to focus the lens. Under these conditions, the magnification of the lens is close to the minimum value possible for the given values of V_3 and V_5, because the $V_1-V_2-V_3$ lens contributes minimally to the total power. To find a value for the magnification close to the maximum possible value, V_4 is set to its minimum value and V_2 set to the greater of V_1 and V_3 and adjusted to focus the lens again.

If the target magnification lies outside these limits, the program halts with a message that the maximum (or minimum) value has a certain value. This value is given as some 5% less than the true maximum as a safeguard against later problems if V_3 is changed. If the target magnification is acceptable, then V_4 is set to a value found by interpolating between its values for the minimum and maximum magnification, and V_2 is adjusted to focus the lens. The magnification is compared to the target value and V_4 is readjusted. Cycles of focusing using V_2 and magnification adjustment using V_4 continue until the target value is met to within 0.1%. This is quite good enough for any practical purpose, but it is noticeable that the dependence of magnification on the ratio V_4/V_2 is often quite weak.

References

[1] Hawkes P W and Kasper E 1989 *Principles of Electron Optics* (London: Academic)

[2] Szilagyi M 1988 *Electron and Ion Optics* (New York and London: Plenum)

[3] Harting E and Read F H 1976 *Electrostatic Lenses* (Amsterdam: Elsevier Scientific Publishing Company)

[4] Picht J 1932 *Ann. Phys., Lpz.* **15** 926–64

[5] Natali S, Di Chio D and Kuyatt C E 1972 *J. Res. NBS* **76A** 27–35

[6] Edwards D 1983 *Rev. Sci. Instrum.* **54** 1729–35

[7] 1958 *British Association Mathematical Tables* vol 6 (Cambridge:Royal Society)

[8] Bonjour P 1979 *Revue Phys. Appl.* **14** 533–40

[9] Renau A, Read F H and Brunt J N 1982 *J. Phys. E: Sci. Instrum.* **15** 347–54

van Hoof H A 1980 *J. Phys. E: Sci. Instrum.* **13** 1081–9

Martinez G and Sancho M 1982 *Am. J. Phys.* **51** 170–4

[10] Renau A and Heddle D W O 1986 *J. Phys. E: Sci. Instrum.* **19** 284–95

[11] Verster J L 1963 *Philips Res. Rep.* **18** 465–605

[12] Hawkes P W 1987 *J. Phys. E: Sci. Instrum.* **20** 234–5

2 of 2